C and Visual Studio

With Introduction to Azure IoT C SDK, Azure Sphere, and Eclipse ThreadX

By

Sean D. Liming and John R. Malin

Copyright

Dedication

To those who like to get down and dirty with real programming.

Table of Contents

Preface

The place where software and hardware meet was a fun part of my college experience. My professional path has focused on higher-level software programming and custom operating systems architecture. My computer engineering background has helped me with my career, but any chance to get into low-level programming and interacting with hardware has been a joy.

I first learned C programming in college using the old Borland C tools, which were not the greatest that I remember. The class was an outlier in the Electrical Engineering department at the time. There was no follow-up or integration into other parts of the curriculum even though the course text book leaned toward electrical engineering examples. In the end, other course programming projects were in assembly. Programming in C was only done as a hobby over the years. Career and job opportunities required Visual Basic and C#, and at one point Java.

With Microsoft addressing a product gap to support Azure with RTOSes, SDKs, and a custom MCU, programming in C has become important again. There is a catch to their approach. Many students in school are learning computer languages like Java, Perl, Ruby, and Python to target AI, web apps, and big data computing. With all the attention on these languages, C language programming in education has kind of disappeared. Many low-level devices still use C with some assembly as the primary programming language. If AI is really going to take off, then all the little devices running C and connecting to the Cloud are going to be important. There is a new opportunity for articles and books on C programming. There are many books that cover the C using GCC and Eclipse, but the MCU solutions from Microsoft can be developed in Visual Studio. Since C is a standard, it doesn't matter what compiler is used. This book is to serve as a simple introduction to C programming using Visual Studio. The end goal is to write C programs for the current MCU offerings from Microsoft and move on to creating C programs using other tools for MCUs, FPGAs, and interacting with Azure.

The focus is on the basics, so anyone reading this book should have a basic understanding of computer programming and how computers work. Please let me know if there is anything that you feel is missing or can be improved upon. You can send suggestion to the contact page: www.annabooks.com.

Sean D. Liming
Rancho Mirage, CA

Acknowledgments

Special thanks to Avnet for providing information on Azure Sphere, as well as, pictures of their starter kit. We would like to thank our family members who have put up with us as we focus on writing another book.

Annabooks

Annabooks provides a unique approach to embedded system services with multiple support levels. Our different offerings include books, articles, training, and project consulting. Current publications and courses have focused on embedded PC architecture and Windows Embedded, which reach a wide audience from Fortune 500 companies to small organizations. We will continue to expand our future services into new technologies and unique topics as they become available.

Books and eBooks
Stater Guide to Windows 10 IoT Enterprise
Java and Eclipse for Computer Science
Open Software Stack for the Intel® Atom™ Processor
Professional's Guide to POS for Windows Runtime
Professional's Guide to POS for .NET
Real-Time Development from Theory to Practice
The PC Handbook

Training Courses
Windows® 10 IoT Enterprise Training Course

Web: www.annabooks.com

1 It all Begins with C

In the past 80 years, there have been 700+ computer programming languages developed. In the last 50 years, the C programming language remains the core programming language for most systems. C is the basis for the UNIX (LINUX) kernel, ported to almost every processor architecture, and is the foundation for other programing languages. C is at the core of our modern computing world. There are many C programming books, but most of them use the GCC. With the development of Azure IoT C SDK and Azure Sphere, the purpose of this book is to get you familiar with programming with C using Visual Studio. This book is intended to be read cover-to-cover. The concepts build from chapter to chapter. This chapter provides some background and sets up the development environment so you can begin writing programs in C right away.

1.1 A Very Brief History

In 1969, Bell Labs was looking to create a programming environment for the DEC PDP-7. The first effort was the UNIX operating system written in the PDP-7 assembly language. A higher-level language was needed to improve operating system development and provide portability. Dennis M. Ritchie started the development of C in the 1970's with the first appearance in 1972, followed by the well-know K&R C in 1978 (Kernighan & Ritchie). There is a whole back story of other programming languages that led to C such as BCPL, B, Fortran, and Algo 68. Dennis wrote a book and paper covering the development of C. In short, the C language was developed on UNIX and then UNIX was subsequently re-written in C. The advantages of C were that it was just above assembly code giving the same capabilities of assembly language without having to write assembly code, and it could be ported to any computer architecture, since C was not tied to UNIX or the PDP-7 directly. C could and was ported to different processor architectures, and eventually was made into an ANSI standard in 1989: ANSI X3J11. There are different compilers available to target different hardware and operating systems. Visual Studio has the C/C++ compiler (MSVC), targets Windows and Linux, and uses the open-source GNU compiler. All these

1

compilers follow the ANSI C standard. Each C standard produced by the governing body is designated as C plus the year of conception. The latest release is C17 (aka C 2017).

1.2   *Future MCU Books and Articles from Annabooks*

Many C Language books have been written in the last 50 years including the first and most important by Brian W. Kernighan and Dennis M. Ritchie (K&R). With so many higher-level languages getting all the market attention, C often gets overlooked. Why are we writing this book? Microsoft has a focus on Azure, and what is going to make Azure successful is the billions of IoT devices that are going to be connecting to the Cloud in the coming decades. The problem is that Microsoft only offers Windows 10 IoT Enterprise, which only supports 64-bit Intel CPUs and 64-bit ARM MCUs from NXP and Qualcomm. Many of the MCU devices being used in systems today are running on ARM 32-bit. Since Windows CE went end-of-life and Windows 10 IoT Core was dead-on-arrival, Microsoft has slowly evolved an ARM 32-bit MCU strategy that includes a system-on-chip, SOC, solution called Azure Sphere, the acquisition of the ThreadX™ RTOS, support for FreeRTOS, and bare metal solutions to connect to Azure. All of these solutions require the C programming language to write programs, but there is a catch to this MCU strategy.

The USA educational system teaches students managed code languages such as Java and scripting languages like Python, and most Windows developers are programming in C#. Most C programming books use Eclipse and GCC to compile the book's exercises. Most tools for MCUs and FPGAs are based on Eclipse. The real kicker is that Azure Sphere applications are built in Visual Studio. Azure IoT C SDK was designed to be portable, and this includes support for developing applications in Visual Studio to run on Windows. Finally, the ThreadX RTOS has a port to Windows, which gets built with Visual Studio. Visual Studio is a very powerful development tool that is used to write applications for the cloud to the mobile device. Since the Visual Studio C/C++ compiler is following the standard, why not learn C using Visual Studio. Since there is little coverage of C programming in Visual Studio and Annabooks will be publishing books and articles on these new MCU offerings from Microsoft, the purpose of this book is to cover C programming in our practical style and avoid having to include a C programming primer in every book in the future. The programs in this book will target the PC platform, which has vast amounts of memory and performance, but we will address the technical and design decisions for small memory-constrained devices. We will not cover every function in the C library but focus on the core principles of C programming that can be used for any programming environment.

1.3 Book Organization and Requirements

If you are not familiar with our practical writing style, it simply means hands-on learning. Each chapter has some computer activity to be performed. Chapter 1 provides a basic introduction to installing Visual Studio and writing your first program. Chapters 2 through 9 cover the basic C programming concepts that would be taught in any Computer Science course. These first 9 chapters are designed to be read from beginning to end. Each chapter builds or adds to the other to weave a cohesive story. Chapter 10 introduces the Azure IoT C SDK which allows you to write programs that send data to the Azure. Chapter 11 introduces writing programs for Azure Sphere. Finally, Chapter 12 covers a unique concept to run ThreadX™ and GuiX™ on the PC platform.

Visual Studio 2022 and the C templates we provide are the only software requirements for chapters 1 through 9. Chapters 10, 11, and 12 will require other tools to be downloaded and installed. Chapter 11 requires an Azure Sphere hardware platform, in which the details are covered in the chapter. With all the background particulars covered, let use jump into our first computer activity.

1.4 Computer Activity 1.1: Installing Visual Studio 2022 and C Project Template

Visual Studio comes with a C/C++ compiler. Although the project templates support the major languages such as C++, C#, VisualBasic, Python, etc., there is no C template available. Since the built-in C/C++ compiler follows the C standard, you can still program with C in Visual Studio. You can take a C++ project and turn it into a C project, but we don't want to have to do this each time. To make development easier, we have created a Basic C template for you to use in all the projects in chapters 1-10.

1. Open a Browser and go to https://visualstudio.microsoft.com/.
2. Visual Studio has grown the support family to support the Mac. Our focus is on Visual Studio running on a Windows PC. Under Visual Studio, there are different editions available. The Community edition is free. You may choose any edition

you like, but the Community edition is all that is needed for this book. Click Community.

3. A VisualStudioSetup.exe file will download. Run the executable.
4. A dialog will appear asking to download the support files for the install wizard, click Continue.
5. After all of the items have been downloaded, the install wizard will appear. Visual Studio is a large development environment capable of supporting different programming languages and types of projects. The wizard provides a selection of options so you can choose what you want to target. In the Workloads section select the following:

• Desktop development with C++
• Visual Studio extension development

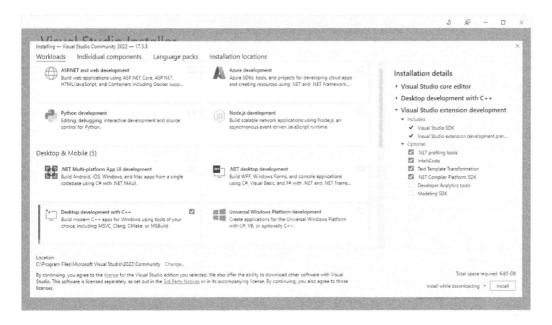

6. Click the Install button.

Note: you can have different versions and editions of Visual Studio installed on your system. Visual Studio Installer will support future updates for all the different versions that are installed.

7. After the installation is complete, the Visual Studio Community edition will launch.
8. Download the support files from the book page on http://www.annabooks.com/Book_CandVS.html.
9. In the Chapter 1 folder is the CLanguageTempInstaller.vsix which contains a C language project template that we will use for this book. Double-click on CLanguageTempInstaller.vsix.
10. A dialog will appear asking to select the product to add. Check the Visual Studio Community edition and click Install.

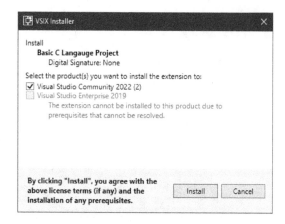

11. Once the template has been installed, click the Close button.

1.5   Computer Activity 1.2: Writing the First Program

Now, we are ready to write our first application.

1. Open Visual Studio.
2. Click on "Create a New Project".
3. You should see the "C Language Project" template at the top of the project list. If not, you can search for the template. Select the "C Application" template, and click Next.

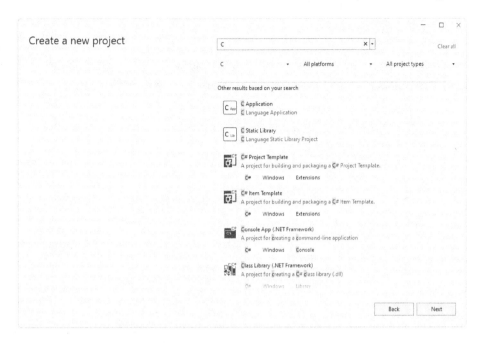

4. Enter CH1-HelloWorld for the Project Name, and provide a project directory.
5. Click the Create button.
6. Visual Studio opens to the editor. In the Solution Explorer on the right, open Source.c. You will see the core C application already set up:

```c
#include <stdio.h>

void main() {

}
```

There are only two types of files supported in C: .c source file and .h header file. Every C application has a main() function. The include makes accessible the standard C IO library APIs available for the program.

7. In the main(), add printf("Hello World\n");

```c
#include <stdio.h>

void main() {
    printf("Hello World\n");
```

7

```
}
```

The printf() function is one of the core functions in stdio.h. Using the .h file, the compiler will be able to pull in the right code from the library for the printf() function.

8. From the menu, select Build->Build Solution. The output window at the bottom should show no errors:

```
Build started...
1>------ Build started: Project: CH1-HelloWorld, Configuration: Debug
x64 ------
1>Source.c
1>CH1-HelloWorld.vcxproj -> E:\C-Apps\CH1-HelloWorld\x64\Debug\CH1-
HelloWorld.exe
========== Build: 1 succeeded, 0 failed, 0 up-to-date, 0 skipped
==========
```

By default, the code is compiled to the ISO C++ 14 standard. You can change to the latest standard via the project properties.

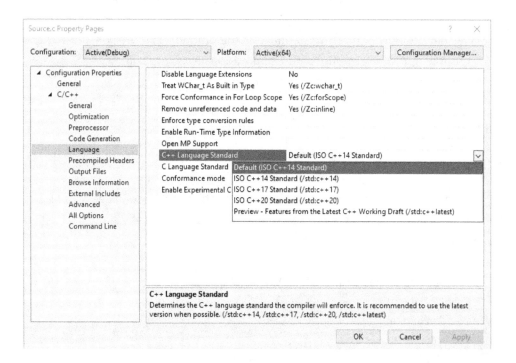

9. Since the application is a console application, we want to see the output without the console closing. Click on the gray space next to line 4 to set a breakpoint.

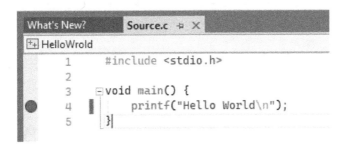

10. Hit F5 on the keyboard to start debugging.
11. The code will hit the breakpoint click on the step-over button or F10. The Hello World message will appear on the consol.
12. Click continue and the application and debug session will come to an end.

1.6 Visual Studio Debugging Features

Visual Studio comes with a rich set of debug features that can help troubleshoot problems with the code. Since C is built on assembly, you can see the code in assembly in the disassembly window.

```
Disassembly  -p  ×  Source.c
Address: main(...)
⌄ Viewing Options
    --- E:\C-Apps\CH1-HelloWorld\CH1-HelloWorld\Source.c --------------------------
    #include <stdio.h>

    void main() {
    00007FF6AD911860  push      rbp
    00007FF6AD911862  push      rdi
    00007FF6AD911863  sub       rsp,0E8h
    00007FF6AD91186A  lea       rbp,[rsp+20h]
    00007FF6AD91186F  lea       rcx,[__0BFE163A_Source@c (07FF6AD921008h)]
    00007FF6AD911876  call      __CheckForDebuggerJustMyCode (07FF6AD91135Ch)

        printf("Hello World\n");
➡ ▷00007FF6AD91187B  lea       rcx,[string "Hello World\n" (07FF6AD919C28h)]
    00007FF6AD911882  call      printf (07FF6AD91118Bh)


    }
    00007FF6AD911887  xor       eax,eax
    00007FF6AD911889  lea       rsp,[rbp+0C8h]
    00007FF6AD911890  pop       rdi
    00007FF6AD911891  pop       rbp
    00007FF6AD911892  ret
    --- No source file ---------------------------------------------------------
    00007FF6AD911893  int       3
    00007FF6AD911894  int       3
```

The CPU registers are also viewable, as well as, the call stack, program memory usage, program performance measurements, events, and Locals.

```
Registers
 RAX = 0000000000000001 RBX = 0000000000000000 RCX = 00007FF6AD921008 RDX = 000001DEF510FFB0 RSI = 0000000000000000
 RDI = 0000000000000000 R8  = 000001DEF5110040 R9  = 000000331FF3FD08 R10 = 0000000000000012 R11 = 000003331FF3FDE
 R12 = 0000000000000000 R13 = 0000000000000000 R14 = 0000000000000000 R15 = 0000000000000000 RIP = 00007FF6AD91187
 RSP = 000000331FF3FD10 RBP = 000000331FF3FD30 EFL = 00000200
121 %  ⌄
Memory 1 | Memory 2 | Memory 3 | Memory 4 | Registers
```

Since all the programs in this book are console programs, we will just run debug applications to simply the learning process. The debug tools will be very helpful when we get to the chapters on pointers and memory management.

1.7 Source Code License and Warranty

The code for the various computer activities can be found on the book page at www.annabooks.com/book_candVS.html. The source code is provided for learning purposes. The source code provided is free to use and is supplied as is without any warranty. The Chapter 10 example uses a sample from Microsoft that is assigned the MIT license; but again, this is free to use so long as you follow the MIT license agreement.

1.8   The Standard C Library

Stdio.h is one of the 29 header files that define the C standard library. The original C library had only 15 header files, but the library has grown over time. Covering all the functions and libraries would make for a very big book. Instead, we will touch on the various libraries and functions that cover the basic programming topics. If you want to dig further into the C library, the details can be found online:

- C standard library - Wikipedia
- The Current C Programming Language Standard - ISO/IEC 9899:2018 (C18) - ANSI Blog

As mentioned earlier, different compilers or build environments will support different library functions. The Visual Studio compiler will support different functions than the GCC compiler. A development environment for an FPGA might include a subset library to address space requirements. You will have to adjust your programs and search the internet for differences or equivalent functions. The general topics in this book are universal for many implementations of C.

Header Name	Description
<assert.h>	Contains the assert macro, used to assist with detecting logical errors and other types of bugs in debugging versions of a program.
<complex.h>	A set of functions for manipulating complex numbers.
<ctype.h>	Defines a set of functions used to classify characters by their types or to convert between upper and lower case in a way that is independent of the used character set (typically ASCII or one of its extensions, although implementations utilizing EBCDIC are also known).
<errno.h>	For testing error codes reported by library functions.
<fenv.h>	Defines a set of functions for controlling the floating-point environment.
<float.h>	Defines macro constants specifying the implementation-specific properties of the floating-point library.

<inttypes.h>	Defines exact-width integer types.
<iso646.h>	Defines several macros that implement alternative ways to express several standard tokens. For programming in ISO 646 variant character sets.
<limits.h>	Defines macro constants specifying the implementation-specific properties of the integer types.
<locale.h>	Defines localization functions.
<math.h>	Defines common mathematical functions.
<setjmp.h>	Declares the macros setjmp and longjmp, which are used for non-local exits.
<signal.h>	Defines signal-handling functions.
<stdalign.h>	For querying and specifying the alignment of objects.
<stdarg.h>	For accessing a varying number of arguments passed to functions.
<stdatomic.h>	For atomic operations on data shared between threads.
<stdbool.h>	Defines a boolean data type.
<stddef.h>	Defines several useful types and macros.
<stdint.h>	Defines exact-width integer types.
<stdio.h>	Defines core input and output functions
<stdlib.h>	Defines numeric conversion functions, pseudo-random numbers generation functions, memory allocation, process control functions
<stdnoreturn.h>	For specifying non-returning functions
<string.h>	Defines string-handling functions
<tgmath.h>	Defines type-generic mathematical functions.
<threads.h>	Defines functions for managing multiple threads, mutexes, and condition variables
<time.h>	Defines date- and time-handling functions
<uchar.h>	Types and functions for manipulating Unicode characters
<wchar.h>	Defines wide-string-handling functions
<wctype.h>	Defines set of functions used to classify wide characters by their types or to convert between upper and lower case

1.9 Summary

Since the C programming language is fundamental to all computing systems, C is a very important programming language to get familiar with. With cloud computing and billions of IoT devices connecting to the Cloud, there is no way to avoid the C programming language.

2 Data Types, Math, and Strings

One of the biggest questions high school students ask all the time: "Why do we have to learn algebra?" Seeing how algebra is applied in the real world is difficult to show when you are first learning it. You get a glimpse of applying algebra in chemistry and physics courses, but it is computer science and programming where you get to see practical uses of algebra. In this chapter, we will see how to perform basic math in C; and then, we will cover the use of strings to make the output more informative and to get information from the user.

Note: Make sure that you have installed Visual Studio and the Basic C Template per Chapter 1.

2.1 Data Types and Math Operators

One of the first lessons in algebra covered different number sets such as natural numbers, whole numbers, integers, real numbers, and complex numbers. The reason why you needed to learn number sets might not have been clear, until now. Computers process 1s and 0s. Writing code with 1s and 0s can be done, but it is not practical. Java and other high-level languages turn human-readable lines of source code and data into 1s and 0s for the computer to read and process. You need to define the type of data in the source code, so the proper encoding to 1s and 0s takes place. Different compilers will support different data types and sizes. The table below provides a list of the common types.

Data Type	Size (Bytes)	Range	Format Specifier
char	1	-128 to 127 or 0 to 255	%c or %s for strings
unsigned char	1	0 to 255	%c
signed char	1	-128 to 127	%c
int	4	2,147,483,648 to 2,147,483,647	%d or %i

unsigned int	4	0 to 4,294,967,295	%u
short	2	-32,768 to 32,767	%hd
unsigned short	2	0 to 65,535	%hu
long	4	-2,147,483,648 to 2,147,483,647	%ld or %li
float	4	1.2E-38 to 3.4E+38	%f
double	8	2.3E-308 to 1.7E+308	%lf
long double	10 to 16		%Lf

Other programming languages use similar data types, but the syntax will not be exactly the same as the C language definitions. For example, C# defines a byte as a value from 0 to 255 and an sbyte as -128 to 127. As you can see from the list, there are several different versions of integers. The larger the maximum value of the integer type, the more bits are needed, and the more memory is used for the data storage. Float and double are used for very large floating-point numbers. Notice that the usage for larger numbers doesn't use a comma (,) as part of the internal storage. If you need/want commas, they need to be explicitly coded into the formatting of the string representation of the data or they need to be added when displayed or printed out.

Now that we can define data, we can perform some operations with the data. In algebra classes, you learned about the order of math operations (PEMDAS). Computer languages follow the same math rules. The table below shows the basic operators including one extra that you might not have seen before.

Operator	Description
()	Parentheses
*	Multiplication
/	Division
%	Modulus / Remainder
+	Addition
-	Subtraction

The modulus operator yields the remainder of an integer division operation. For example, if 7 and 3 are integers, 7 % 3 is 1 because 7 / 3 = 2 with a remainder of 1.

Because programs can become very large with lots of typing, programmers like to use shortcuts whenever possible. In addition to these basic math operators, C includes a set of increment, decrement, and math-assignment operators.

Operator	Description	Usage	Equivalent
++	Pre or post Increment	Pre: ++x Post: x++	x++ is the same as x = x + 1
--	Pre or post decrement	Pre: --x Post: x--	y++ is the same as y = y - 1
*=	Multiplication assignment	x*= value	z *= 5 is the same as z = z * 5
/=	Division assignment	y /= value	j /= 5 is the same as j = j / 5
%=	Modulus assignment	i %= value	k %= 5 is the same as k = k % 5
+=	Addition assignment	j += value	w += 5 is the same as w = w + 5
-=	Subtraction assignment	k -= value	i -= 5 is the same as i = i - 5

The increment and decrement operators increment or decrement the numeric value by 1. For example:

```
int x = 5;
 x++;
```

The equal sign (=) is known as the assignment operator. The statement int x = 5 means that x is created as data type int and is assigned the value of 5. After x++ runs, the value of x is now 6. The increment can also be viewed as int x = x + 1. The pre- and post-operations are important. For example:

```
int x = 10;
 int y = 5;
 int j = ++x - 2;
 int k = y-- + 3
 int z = x * y;
```

The x variable is incremented before the subtraction operation; thus, j equals the value 9. The y variable is decremented after the addition operation; thus, k equals the value 8. Both x and y have new values because of the increment and decrement operations in the previous statements, thus z equals the value 44 (11 * 4). The shorthand will become useful when we talk about iteration in the next chapter. The math assignment operators are similar. For example:

```
int y = 7;
 y += 4;
```

The assignment can be viewed as y = y + 4. The value of y is 11. Now let's use Visual Studio to test data types and the math operators.

2.1.1 Computer Activity 2.1: Data Type Test

Create a new C Project in Visual Studio and name the project CH2-Data-Types. Enter the following code:

```c
#include <stdio.h>

void main() {

        int x = 5;
        double y = 2.5;
        char c = 'g';

        printf("x = %d\n", x);
        printf("y = %lf\n", y);
        printf("c = %c\n", c);
        printf("Finished");

}
```

Set a breakpoint on the last printf() function and run the program. The output will be:

x = 5

y = 2.500000

c = g

Click continue to finish the program.

2.1.2 Computer Activity 2.2: Testing Math

Create a new C project in Visual Studio called CH2-Math-Test. Enter the following code:

```
1.      #include <stdio.h>
2.
3.      void main() {
4.              int x = 9;
5.              int y = 2;
6.              int i = 100;
7.              double j = 1.5;
8.
9.              x += y;
10.             j++;
11.             y = 5 + x * 3;
12.             i %= y;
13.             j += i--;
14.             y -= j;
15.
16.             printf("x= %d\n", x);
17.             printf("y= %d\n", y);
18.             printf("i= %d\n", i);
19.             printf("j= %lf\n", j);
20.
21.             printf("Finished");
22.     }
```

Set a breakpoint at the last printf() function and run the program. The output will be as follows:

x= 11
y= 11
i= 23
j= 26.500000

2.1.3   Mixing Data Types and Casting

In the last exercise, notice that j is a double. You need to be careful when mixing data types. The result from the program at line 13 was correct, since j can take the result from an integer operation. Because of the type mismatch in line 14, y -= j, since y is an integer and J is a double, y will be promoted to a double before the subtraction is done, and then the result will be converted to integer when the result of the subtraction is stored back in y. We can force a data type change if we use the cast (data type) operator. To change the data type of j to an integer before the subtraction is done, in line 14, substitute the following:

```
y -= (int)j;
```

Run the program and you will notice that the result of y is different from the result you obtained when you did not cast J to an integer. The output is as follows:

x= 11

y= 12

i= 23

j= 26.500000

Either way, the result of y is an integer, but the end value of that integer is determined by the variable types that are in effect at the time of the mathematical operation; in this case, the subtraction. The cast operation (int) temporarily treats the double as an integer when this line is executed. The data type of j does not change. When a double is cast to int, anything after the decimal is dropped. It is not rounded to an integer. It is truncated to an integer. Another area to watch out for is integer division. The code listing below has an example:

```c
#include <stdio.h>

void main() {

        int y = 11;
        int z = 2;
        int k = 4;
        int w = (y + z) / k;

        printf("w = %d\n", w);

        printf("Finished");
}
```

The real value of w is a decimal value of 3.25. Since w is an integer data type, 3 is printed to the console, and the .25 value is dropped. Remember: computers do what they are told, and if you accidentally mix types you might run into unexpected results. The proper solution, if the fractional part of the division is significant and needs to be stored and printed out, is to make w a double and to cast y, z, and k as double, as shown here:

```c
#include <stdio.h>
```

```
void main() {

        int y = 11;
        int z = 2;
        int k = 4;
        //int w = (y + z) / k;
        double w = ((double)y + (double)z) / (double)k;

        printf("w = %lf\n", w);

        printf("Finished");
}
```

2.2 Math Functions

Complex operations such as square roots, exponent, powers, and trigonometry functions are already available as methods in the C <math.h> library. The table below is a short list of some of the available functions in the <math.h> library.

Math Function	Description
abs(int x)	Get the absolute value of x. Returns value of double or int
ceil(double x)	Rounds up to the next integer. Returns value of double
cos(double x)	Gets the cosine of x. Returns radians as double.
exp(double x)	Returns e^x as double
floor(double x)	Rounds down to the next integer. Returns value of double
log(double x)	Returns natural logarithm (to base e)
fmax(double or int x, double or int y)	Returns the larger of the two values
fmin(double or int x, double or int y)	Returns the smaller of the two values
pow(double x, double y)	Returns the value of x^y Return value of double.
rand()	Returns a generated pseudo-random number between 0 and RAND_MAX
round(double x)	Rounds the value of x up to the next integer if >= x.5. Rounds the value down if < x.5. Returns value is a long
sin(double x)	Returns the sine of y. Returns radians as double
sqrt(double x)	Returns the square root of x. Returns a double
tan(double x)	Returns the tangent of x. Returns radians as double

There are over 120 available functions defined in <math.h>, and the details can be found https://en.wikipedia.org/wiki/C_mathematical_functions#Overview_of_functions. Most of the functions return double as the data type, so be careful when assigning values.

2.2.1   Computer Activity 2.3: Math Test 2

Let's do a simple test to see these methods in action. Create a new C project in Visual Studio called CH2-Math-Test2. Enter the following code:

```
#include <stdio.h>
#include <math.h>

void main() {
```

```
        double x = 2.6;
        int y = 7;

        int j = abs(y);
        printf("j = %d\n", j);

        double k = ceil(x);
        printf("k = %lf\n", k);

        double t = floor(x);
        printf("t = %lf\n", t);

        double maxv = fmax(x, y);
        printf("maxv = %lf\n", maxv);

        double minv = fmin(x, y);
        printf("minv = %lf\n", minv);

        double z = round(x);
        printf("z = %lf\n", z);

        double r = rand();
        printf("r = %lf\n", r);

        printf("Finished");

}
```

Set a breakpoint at the last printf function and run the program. The values x and y are passed to the different methods. Notice that the Round() method rounds the value up to 3. Change x to 2.4 and run the program again. Notice that the result is now 2. You can combine these math methods and the basic math operators from the last section to turn algebraic equations into C program statements. For example, the following

$$ y = \left(x + \frac{\sqrt{b}}{5} \right)^3 $$

becomes the following statement in a Java program:

```
    double y = pow((x+(sqrt(b)/5)), 3);
```

It takes some practice to see and write equations as single-line statements in a program. A typical error occurs if a parenthesis is missed. If you miss a parenthesis, Visual Studio will let you know immediately.

2.3   Variables and Constants

To this point, we have used single letters as variables such as x, y, j, etc. These variables are known as local variables as they are declared within the {} of the function and exist only in that function. C defines several variable types that we will cover in later chapters. Variables can also be words like 'counter', 'volume', or 'led1'. It is common practice for variables to start with a lower case. There are some reserved keywords in C that you cannot use as variables, such as int, public, static, while, for, import, etc. These reserved words are used as directives for the C language compiler. Visual Studio will warn or generate errors if you use these keywords. Sometimes you might be writing an algorithm for an equation that has a constant value. For example:

$$e = mc^2$$

The value for c is the speed of light in a vacuum or 299,792,458 meters per second. Having to enter the value each and every time would be a bit cumbersome. You could assign the value to 'c' as we have done in the previous exercises and examples, but the value of 'c' could be overwritten if you forget that it is a constant. The keyword "const" is used to prevent the value from changing:

```
const double c = 299792458;
double e = m * c * c;
```

The final declaration means the value cannot change anywhere in the program. Visual Studio will flag an error if 'c' is assigned another value or changed in the program. In this case, 'c' was lower case, but in practice, it is common practice to capitalize constant values so they stand out from variables that can change. For example:

```
const double MAXHEIGHT = 42;
```

2.4 Bit Manipulation

When programming for MCU or FPGA devices, especially for systems that do not have an OS, it becomes necessary to be able to read, manipulate, and write individual bits. These are usually the bits of control and data registers for the devices like a UART or network controller. C provides the following bit manipulation operators:

Bitwise Logical Operator	Operation
&	Bitwise AND
|	Bitwise Inclusive OR
^	Bitwise Exclusive OR
<<	Left Shift
>>	Right Shift
~	One's Complement

These operators may be applied to integer type variables but not to float or double type variables.

The bitwise AND operator, is typically used to mask or clear bits. If we have an 8-bit integer, Test, that has a value of 0x87, but we want to only look at the lower 4 bits, we could AND it with 0x0F and we would have:

Test = 0x87

Test & 0x0F = 0x07

The bitwise OR operator is typically used to set bits. The bitwise exclusive OR operator is used to selectively set bits.

As their name implies the left and right shift operators shift the bits of a variable left or right by the number of bits indicated.

The one's complement operator inverts the bits, i.e., 1's become 0's and 0's become 1's.

2.4.1   Computer Activity 2.4 Bit Fiddling

Let's try some of these bit fiddling operators and see how they work. Create a new C project in Visual Studio called CH2-Bit-Fiddling. Enter the following code:

```c
#include <stdio.h>

int main() {
        unsigned __int8 MockRegister = 0x07;
        unsigned __int8 MaskReg = 0x80;
        unsigned __int8 BinaryMask = 0x80;

        printf("CH2 Bit Fiddling Computer Activity\n\n");

        printf("Sizeof MockRegister is %d bytes and initial value = 0x%02x\n", (int)sizeof(MockRegister), MockRegister);

        //Apply mask register
        MockRegister |= MaskReg;
        printf("New MockRegister value = 0x%02x\n", MockRegister);

        //Display Mockregister as a binary number using & and >>


        //->1 high order bit
        printf("MockRegister in binary is: ");
        if ((MockRegister & BinaryMask) == 0)
        {
                printf("0");
        }
        else
        {
                printf("1");
        }
        BinaryMask >>= 1;

        //->2
        if ((MockRegister & BinaryMask) == 0)
        {
                printf("0");
        }
        else
        {
                printf("1");
        }

        //->3
        BinaryMask >>= 1;

        if ((MockRegister & BinaryMask) == 0)
        {
```

```
        printf("0");
}
else
{
        printf("1");
}

//->4

BinaryMask >>= 1;

if ((MockRegister & BinaryMask) == 0)
{
        printf("0");
}
else
{
        printf("1");
}

//==next nibble
//
        //->5
BinaryMask >>= 1;

if ((MockRegister & BinaryMask) == 0)
{
        printf("0");
}
else
{
        printf("1");
}
BinaryMask >>= 1;

//->6
if ((MockRegister & BinaryMask) == 0)
{
        printf("0");
}
else
{
        printf("1");
}

//->7
BinaryMask >>= 1;
```

```
        if ((MockRegister & BinaryMask) == 0)
        {
                printf("0");
        }
        else
        {
                printf("1");
        }

        //->8

        BinaryMask >>= 1;

        if ((MockRegister & BinaryMask) == 0)
        {
                printf("0");
        }
        else
        {
                printf("1");
        }

        printf("\n");

}
```

Set a breakpoint at the last printf function and run the program. You will see the following output:

CH2 Bit Fiddling Computer Activity

Sizeof MockRegister is 1 byte and the initial value = 0x07
New MockRegister value = 0x87
MockRegister in binary is: 10000111

You will see that a mock register 8-bits wide is created. In the real world, this would probably be a hardware register that would be mapped via a pointer (more on pointers later). We will assume that this register has a current status value of 0x07. The MockRegister will be read and output via a printf command. In that print statement, we will verify the size of our MockRegister and show the current value.

Next, we will set a bit in that register to interact with our mock device. In the real world this might be sending the hardware a command to do something. This will be done using the bitwise OR with the MaskReg variable that is the same size and type as MockRegister. ORing the MockRegister with the MaskRegister will set the high order bit in the register.

Finally, we will verify the new value of the MockRegister, and then we will output the register's value as a binary number. To do this we will use the BinaryMask variable, again the same type and size as the MockRegister, and we will test each bit using a bitwise AND of the MockRegister with the BinaryMask variable. As each bit is tested and output via a printf command, the BinaryMask with be changed to test the next bit using the right shift, >>, operator. This process of testing each bit and shifting the test mask is continued until all the bits have been tested and displayed.

You can see that repeating the code to test each bit is rather tedious and prone to making typo errors. When we get to the next chapter, we will see how to use a program loop to accomplish the same results in a more compact and less error-prone way.

2.5 Random Number Generator

The <math.h> or <stdlib.h> libraries includes a rand() method that generates a pseudo-random integral number between 0 and RAND_MAX (32767 in this library's implementation). A random number generator is a fun, useful feature that can be used where random events or values are needed, such as the rolling of the dice or dealing a deck of cards. Quality assurance and test software programs take full advantage of a random number generator to simulate user input, sensor data, traffic flow, weather, etc. There are two basic random number functions in these libraries: srand() and rand().

Random Function	Description
srand(unsigned x)	Sets the seed of the random generator
rand()	Returns a pseudo-random number between 0 and RAND_MAX

If you re-run Computer Activity 2.3: Math Test 2 a few times, you will notice that the random number is the same value. That is because the rand() function's algorithm requires a seed value to start the calculation. If no see value is provided, the rand() function defaults to a seed of 1 when it is first called. To override the default seed value the srand() function is used. It seeds the rand() function with a new value to base the random numbers on. If we use a different seed number, a new pseudo-random number sequence gets generated, but the random number sequence will again be the same on each re-run of the application if the same seed number is always used. This is why the random generator is called a pseudo-random generator. The only way to get a true random number is to seed the random number generator with a different value every time the program is run. The best way to seed the random number generator with a random number is to use the current time function, time(), which returns the number of seconds elapsed since midnight, January 1, 1970, in this particular implementation. This is implementation-specific, if you want to translate the time value to absolute time and date. When used as a seed number, however, we only need to know that the number of seconds that we get back, each time we read them, will always be different; because time keeps advancing and each time value returned will be unique. Note: if you call time() multiple times in 1 second you cannot be assured of different results each time. You might ask why we don't simply use the time value, itself, as the random number. That is because the relationship between reads of the system time is not random, but it does guarantee a different seed number for the pseudo-random number which in turn provides a unique series of random numbers.

2.5.1 Computer Activity 2.5: Random Number Test

Let's test some ideas around generating random numbers. Create a new C application in Visual Studio called CH2-Random-Test. Enter the following code:

```c
#include <stdio.h>
#include <math.h>
#include <time.h>
#include <stdlib.h>

int main() {
        int d1, d2, d3, d4, d5;
        int r1, r2, r3, r4, r5;

        srand(51);
        d1 = rand();
        d2 = rand();
        d3 = rand();
        d4 = rand();
        d5 = rand();
```

```
        printf("d1 = %d\nd2 = %d\nd3 = %d\nd4 = %d\nd5 = %d\n\n", d1,
d2, d3, d4, d5);

        srand((unsigned int)time(NULL));
        r1 = rand();
        r2 = rand();
        r3 = rand();
        r4 = rand();
        r5 = rand();

        printf("r1 = %d\nr2 = %d\nr3 = %d\nr4 = %d\nr5 = %d\n\n", r1,
r2, r3, r4, r5);

        printf("\n");


}
```

The <time.h> library is included so the application can use the time() function to seed the random number generator. Set a breakpoint on the last printf function and run the program a few times. The output the first time you run the program should look similar to this:

First run:
d1 = 205
d2 = 31600
d3 = 14807
d4 = 17250
d5 = 10610

r1 = 2782
r2 = 28866
r3 = 23533
r4 = 8473
r5 = 26103

Second run:
d1 = 205
d2 = 31600
d3 = 14807

d4 = 17250
d5 = 10610

r1 = 3109
r2 = 22362
r3 = 7710
r4 = 22741
r5 = 8985

Third run:
d1 = 205
d2 = 31600
d3 = 14807
d4 = 17250
d5 = 10610

r1 = 3292
r2 = 1681
r3 = 25064
r4 = 26798
r5 = 29545

Notice that the values of the d_n series of random numbers stay the same on each run because the seed number never changes. That is why rand() is a pseudo-random number generator. The values of the r_n series change for each run, because the time is used to provide a unique seed number for each run. We will return to the rand() function in later chapters.

2.5.2   Computer Activity 2.6: Dice
Games are a good application for the random number generator. Create a new C program in Visual Studio called CH2-Dice. Enter the following code:

```
#include <stdio.h>
#include <math.h>
#include <time.h>
#include <stdlib.h>

void main() {
```

```
    int d1, d2, d3, d4, d5;

    srand((unsigned int)time(NULL));

    d1 = rand() % 6 +1;
    d2 = rand() % 6 +1 ;
    d3 = rand() % 6 +1;
    d4 = rand() % 6 +1;
    d5 = rand() % 6 +1;

    printf("d1 = %d\n", d1);
    printf("d2 = %d\n", d2);
    printf("d3 = %d\n", d3);
    printf("d4 = %d\n", d4);
    printf("d5 = %d\n", d5);

    printf("Finished");

}
```

The application will get the random values of 5 dice. The modulus operation (%6) produces a random number between 0 and 5 from the random numbers created by the random number generator, rand(). The addition of 1 results in a random number between 1 and 6, the values that appear on a six-sided die. Set a breakpoint on the last printf() function and run the application a few times to see the random dice results.

2.6   Strings

The printf() function has been demonstrated to output strings in quotation marks, but what about string variables? In higher-level languages alike Java and C#, there is a whole class with methods dedicated to strings. For C, the 'char' data type is all there is for strings. The char is 1 byte in length, which implies only 1 character. To accommodate multiple characters, a character array is used. Arrays and Pointers will be discussed in future chapters, but this strings discussion is a look-ahead. Here is an example of creating a string:

```
    char myString[81] = "Quick Brown Fox";
```

The <strings.h> library provides a number of functions to manipulate strings such as copy strings, concatenate strings, find bytes in a string, and conversation to numbers.

Warning!: There are some functions that are considered unsafe and susceptible to attacks. The functions strcpy() and strcat() in <string.h> are a couple of examples. The functions are a source of buffer overflow vulnerabilities. A good compiler, like the one in Visual Studio, will error on the build if a program tries to use one of these functions. There are safe alternatives to these functions that have been developed. For strcpy(), there is strncpy_s() and for strcat(), there is strncat_s(). Different compilers support different functions. The Visual Studio compiler will have different functions than the GCC compiler.

2.6.1 Computer Activity 2.7: Working with Strings

Create a new C program in Visual Studio called CH2-String-Test. Enter the following code:

```
#include <stdio.h>
#include <string.h>

void main() {

        char myString[100] = "Quick Brown Fox";
        printf("%s\n", myString);

        char myStr1[100] = "Today is ";
        char myStr2[100] = "great day to learn C";

        int errorHandle =  strncat_s(myStr1, 100, myStr2, 20);

        printf("%s\n", myStr1);

        char inputNum[10] = "42";
        int i = atoi(inputNum);
        i++;
        printf("i = %d", i);

        printf("Finished");
}
```

Set a breakpoint on the last printf() function call and run the program. The first string is simply printed to the screen. The %s format specifier prints out the whole string. The next two strings demonstrate the strncat_s() function to concatenate the strings. The format for strncat_s() functions is as follows:

strncat_s(destination buffer, size of the buffer, source buffer, how many characters to copy).

The program copied 20 characters of myStr2 into the available bytes in myStr1. Change the characters to copy to 10, and you will see that the full message has been truncated. The last string gets converted to an integer, incremented, and the result is sent to the standard console output. There are so many string functions and safer versions of the string functions that we cannot cover them all. Please search online for more information on string functions and safe handling of strings.

2.7 User Input and Output

We have been using the printf() function to output information to the standard console output. The printf() function is part of the <stdio.h> library. The format specifiers, such as %d, are placed within the string quotes so that variable data to be printed to the standard console output formatted to our liking. You will notice that \n has also been included. The \n is known as an escape sequence that causes a newline sequence (carriage return – line feed derived from the old typewriter days) to be output, so the cursor goes to the next line and the beginning of that line in the standard console. There are several other escape sequence characters that can be used to format the output. The table below lists the other escape sequences:

Escape Sequence	Description	ACII
\t	Inserts a tab	HT
\b	Inserts a backspace	BS
\n	Inserts a newline character	LF
\r	Inserts a carriage return	CR
\f	Insert a form feed	FF
\'	Insert a single quote character	'
\"	Insert a double quote character	"
\\	Insert a backslash character	\

The scanf() function is used to get user input from the standard input console, but like the strcat() function, scanf() has a vulnerability and is replaced with scanf_s() in the Visual Studio compiler. Let's dive into an example and see how this works.

2.7.1 Computer Activity 2.8: User Input/Output

Create a new C program in Visual Studio called CH2-User-IO. Enter the following code:

```c
#include <stdio.h>

void main() {

        char userInput[50];
        printf("Enter a string and hit Enter...\n");
        scanf_s("%49s", userInput, 50);
        printf("The user entered: %s\n", userInput);

        printf("Finished\n");

}
```

Set a breakpoint at the last printf() function and run the program. When asked to enter a string, enter the following:

This is a test.

Once you hit enter you will notice that the "This" word is part of the output as the rest of the string appears to be lost. The rest of the string is in the buffer waiting to be scanned in. scanf_s() requires a loop to capture all the characters in the input buffer string. The alternative input function that gets the whole buffer is gets_s(). Replace the code with the following:

```c
#include <stdio.h>

void main() {

        printf("Enter a string had hit Enter.\n");
        char userInput2[50];
        gets_s(userInput2, 50);
        printf("The user entered: %s\n", userInput2);


        printf("Finished\n");

}
```

If you run the program this time, the whole input string will be output to the console. The counterpart function that goes with gets_s() is puts(). The puts() function simply outputs a buffer string without any formatting like printf(). For example:

```
puts(userInput2);
```

There are two other character input and output functions. They are getchar() and putchar(). As the function name implies these two functions simply get a 1-byte character or output a 1-byte character respectively. The code listing below provides a demonstrates of these functions.

```
#include <stdio.h>

void main() {

        char ch = 'a';
        putchar(ch);
        printf("\n");
        printf("Enter a character and hit enter\n");
        ch = getchar();
        putchar(ch);

        printf("Finished");
}
```

Notice that we are using the single ' ' quotes when ch is first created. The single quotes are for characters.

2.8   Comments in Code

You may have to share your code with others or go back to your code many years later. You might not remember what and why you wrote the code the way you did, so a good coding practice is to add comments to your code, especially if it is not obvious why the code was written the way it was. Adding comments that describe what your algorithm is doing helps the reader understand the code faster and provides the recall of that creative moment you were having years ago when you originally wrote the code. Comments are added in two ways. The first is a single-line comment that is preceded with two forward slashes //:

```
//This is a comment
```

C treats the "//" and all text that follows until the end of the line as a comment and does not try to process it as code. In Visual Studio and most code editors, the single-line comment is in the color green. If you need to write several lines of text, you can use the block comment that starts with /* and ends with */. In Visual Studio as soon as you enter /**/, you can write as many lines as you like between the *'s.

```
/* this is a comment block
that spans multiple lines
in the program */
```

Visual Studio has buttons in the tool bar to comment out and uncomment a number of lines in a program. There are also the shortcuts CTRL+K, CTRL+C, and CTRL+U that you can use. You can take advantage of this feature when testing different blocks of code. Simply select a number of lines of code and through the tool bar or shortcuts select "comment out the selected lines", or select a number of // commented lines and through the tool bar or shortcuts select "uncomment the selected lines".

Note: You cannot nest block comments, i.e. you cannot have a comment block within a comment block. That will cause an error. You can nest single-line comments inside of a block comment, however.

2.9   *Summary*

All the algebra you learned in school comes back in a big way when you take on software development. The C programming language data types, math functions, and Strings provide the foundation for all the programs and algorithms that will be presented throughout the book. The math functions provide a simple set of tools for basic calculations. More complex algorithms can be created.. Being able to process complex functions or determine how to process the data based on specific inputs is what computers do best; and as you build your calculation algorithms, you will recognize your high school algebra coming to life. This brings us to the next chapter on program flow.

3 Controlling the Program Flow and Iteration

Although Artificial Intelligence (AI) has come a long way since the 1960s, decision-making in programs has been left up to the programmer. A program can change the program flow based on a set of conditions. The program can go in a different direction or loop until a condition is met. The goal of this chapter is to cover program flow control and the concept of iteration.

3.1 Flowchart Diagram

Program flow/conditional branching statements are used to selectively steer the program execution based on the instantaneous state of the program. An old tool called a flow control diagram or flowchart can help draw a picture of the algorithm being developed. In English class, diagramming sentences was a visual aid to help better understand sentence structure. A flowchart provides a visual aid to see the program flow before the code is written. Applications like Microsoft's Visio were created to help generate flowcharts. Below are the basic elements of a program flowchart. We will use these elements to diagram the different flow control statements as a visual aid for discussion.

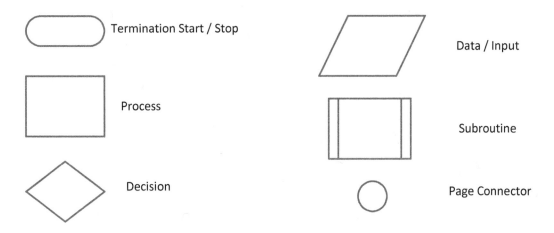

Termination Start / Stop

Data / Input

Process

Subroutine

Decision

Page Connector

Flowcharts are useful tools when you first start programming. In practice, applications are getting more complex and multithreaded, and flowcharts are better used in team programming from a much higher level.

3.2 If-Else Statement, Relational Operators, and Boolean Expressions

After this long introduction, the big decision statement in C is the 'if' statement.

If a condition is true, then perform this action.

The if-statement tests a condition for a Boolean result, which is either true or false. If you want to add some actions to perform when the condition is false, you can add the optional 'else' statement.

If a condition is true, then perform this action, else perform this other action.

Definition: *Condition* – A mathematical statement that resolves to a logical true or false.

Some languages call this the if-then-else statement, and the 'then'- is actually written in the code. The developers for C removed the need for the extra text. The following code listing is a simple program:

```c
#include <stdio.h>

void main() {

    int w = 0;

    if (w == 0)
    {
        printf("W is zero");
    }
    else
    {
        printf("W is not zero");
    }

}
```

The program tests to see if w is equal to zero. If the condition is true, then "W is zero" is printed to the console. If the condition is false, then "W is not zero" is printed to the console. Notice that w = 0 and w == 0 are two different meanings. The single "=" is the assignment operator. The number 0 is assigned to w. The double "==" is the comparison test to see if w equals 0. Many programmers make the basic mistake of confusing assignment with equals so watch out for this in your programs. Also, notice that there is no semicolon ';' after the if- and else- statements. This is because there is a block of code to follow. The figure below shows what the flowchart for the program looks like:

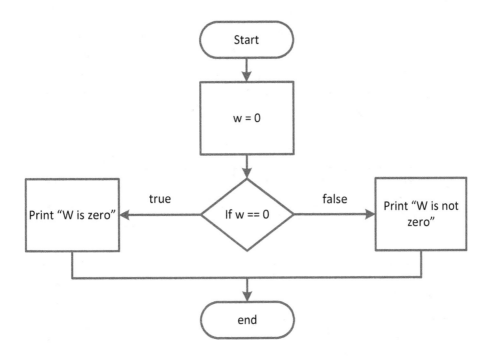

Looking at the flowchart, the if-statement can take the program in two directions. The picture looks like an upside-down tree. You will also hear programmers say the if-statement is a branch statement. The double equals '==' is a Relation Operator. The if-statement tests to see how two objects relate to each other. There are other Relational Operators such as less-than '<' and greater-than '>'. The table below lists the other Relational Operators.

Relational Operator	Description
==	equal-to
<	less-than
>	greater-than
<=	less-than-or-equal-to
>=	greater-than-or-equal-to
!=	Not-equal

What if multiple conditions need to be tested? You can handle these situations using three Boolean expressions: AND (&&), OR (||), and NOT (!). For the AND (&&), all conditions must be true for the if-statement to be true; otherwise, the condition is false. For the OR (||), only one condition must be true for the if-statement to be true; otherwise, the condition is false. The NOT (!) operator reverses the logic from true to false or false to true. For two conditions, the three Boolean expressions can be represented by these truth tables:

AND logic for 2 conditions:

If Condition 1	If Condition 2	Then Condition 1 && Condition 2
False	False	False
False	True	False
True	False	False
True	True	True

OR logic for 2 conditions:

If Condition 1	If Condition 2	Then Condition 1 || Condition 2
False	False	False
False	True	True
True	False	True
True	True	True

NOT logic for 2 conditions:

Condition 1	! Condition1
True	False
False	True

3.2.1 Computer Activity 3.1 – Boolean Test

Create a new C project in Visual Studio Called CH3-Boolean-Test. Enter the following code:

```c
#include <stdio.h>
#include <stdbool.h>

void main() {

    bool x = true;
    bool y = false;
    bool z = true;

    if (x == true && y == false)
    {
        printf("The first condition is true\n");
    }
    else {
        printf("The first condition is false\n");
    }

    if (x == true || y == true)
    {
        printf("The second condition is true\n");
    }
    else {
        printf("The first condition is false\n");
    }

    if (!z)
    {
        printf("The third condition is true\n");
    }
    else {
        printf("The third condition is false\n");
    }
}
```

Set a breakpoint at the first if statement, and run the program. You will notice that each if-else statement uses {} to block the code. Using the {} to define the code block is always a good programming practice. It aids both in management of variables in the code blocks and readability of complex conditional statements. The result of the program is that the first two conditions are true and the last one is false. Adding the <stdbool.h> library adds the bool data type to the program. The program creates three Boolean variables. The first condition is true since x equals true and y equals false, because y equals false doesn't mean the condition fails. The condition that 'y equals false' is a true logical statement. By ANDing both true statements, the full condition is true. The second if-statement condition is also true. The condition that 'y equals true' is false. Since we are ORing the two conditions, the 'x equals true' test is logically true, thus the full condition is true. The final if-statement demonstrates the NOT. The variable z is assigned true. The NOT changes the conditional test to false, and the result is false. Combinations of Boolean tests can be made using parenthesis (), which can make for some complex programming. For example, if we added another condition to the above program:

if (((x == true && !y == true) && (!y == z)) || (!x == z))

The result of this statement would be true. Chapter 2 discussed the order of operation, which also applies to Boolean logic. With complex Boolean expressions, you may want to optimize code for readability. One such solution is using a couple of transformation rules known as De Morgan's Laws:

!(x && y) is the same as !x || !y
!(x || y) is the same as !x && !y

The not (!) operator changes the AND to an OR and an OR to an AND. Also, the not operator changes Relational Operators. Greater-than becomes less-than and less-than becomes greater-than. Here is an example:

!((x >= k) && !(x < y))

Transferring the not (!) through the expression

!(x >= k) || (x < y)

Doing one more transfer, the final expression is

(x < k) || (x < y)

3.2.2   Computer Activity 3.2 – Coin Flip

With conditional branching, we can combine the if-statement with the math and the strings from the previous chapters to develop some creative applications. For this project, you will create a simple coin flip game. The program will ask the user to enter "h" for heads and "t" for tails. The Boolean random generator will act as the coin flip to generate a true for heads or false for tails. The following logic table, Table 3.1, shows that there are 4 possible outcomes. Using the if-statement, the program will print the results.

User Input	Coin Toss	Result
Heads	Heads	Win
Heads	Tails	Loss
Tails	Heads	Loss
Tails	Tails	Win

Table 3.1 Coin Flip Outcomes

Create a new C project in Visual Studio called CH3-CoinFlip. Enter the following code:

```c
#include <stdio.h>
#include <time.h>
#include <math.h>
#include <stdlib.h>


void main() {


    char userSelection;
    int tossResult;

    printf("Enter h for heads or t for tails\n");

    userSelection = getchar();

    if (userSelection == 't') {
            printf("You chose tails\n");
    }
    else
    {
            printf("You chose heads\n");
    }

    srand((unsigned int)time(NULL));
    tossResult = rand()%2;
```

```
    if (tossResult == 0) {
        printf("The coin toss result is tails\n");
    }
    else {
        printf("The coin toss result is heads\n");
    }

}
```

Run the program a few times and enter h and t to see the results. You have now created your first computer game.

3.3   Nested If-Else

An if-else statement can be put into the block of another if-statement. This is called a nested if. The code listing below is an example:

```
#include <stdio.h>

void main() {

    int g = 4;
    int t = 7;
    double n = 8.0;

    if (g < t)
    {
        if (n < t)
        {
            printf("n is less than t, and g is less then t\n");
        }
        else
        {
            printf("t is less than n, and g is less then t\n");
        }
    }

}
```

If the first condition is true, the second condition is tested. If the first condition is false, n < t is never tested. Multiply nested if-statements can be confusing. It is important to indent your code so that others can follow the program flow. The figure below is the flowchart for the above code listing.

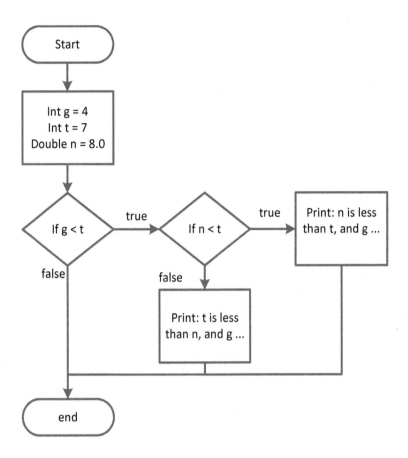

3.4 If-Else-If Ladder

For a series of conditions to test, you can use an if-else-if ladder. If the first condition is false, it tests the next condition. If the second condition is false, the next condition is tested, and so on. A final optional else-statement can be set up as the default condition. The code listing below is an example of this construct.

```c
#include <stdio.h>
#include <string.h>

void main() {

        char str[10] = "car";

        if (strcmp(str, "truck") == 0)
        {
```

```
        printf("I am taking the truck out today\n");
    }
    else if (strcmp(str, "bike") == 0)
    {
        printf("I am taking the bike out today\n");
    }
    else if (strcmp(str, "car") == 0)
    {
        printf("I am taking the car out today\n");
    }
    else
    {
        printf("I am staying home\n");
    }
}
```

We defined a string and assigned it "car". The if-else-if ladder has a condition to test what value str matches. Preforming the test is the strcmp() function from <string.h> library. The program will find the match for "car" and print the string. The following is the flow chart:

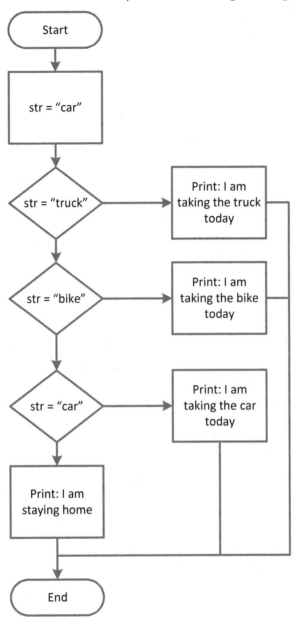

3.4.1 Computer Activity 3.3 – Coin Flip Enhanced

Using the if-else-if ladder, let's modify the coin flip program to provide a different output response for the user. Comment out the original print results:

```c
#include <stdio.h>
#include <time.h>
#include <math.h>
#include <stdlib.h>


void main() {


    char userSelection;
    int tossResult;

    printf("Enter h for heads or t for tails\n");

    userSelection = getchar();

    if (userSelection == 't') {
            printf("You chose tails\n");
    }
    else
    {
            printf("You chose heads\n");
    }

    srand((unsigned int)time(NULL));
    tossResult = rand()%2;

    //if (tossResult == 0) {
    //      printf("The coin toss result is tails\n");
    //}
    //else {
    //      printf("The coin toss result is heads\n");
    //}

    if((tossResult == 0) && (userSelection =='t'))
    {
            printf("The coin is tails. You Win!");
    }
    else if ((tossResult == 1) && (userSelection == 'h'))
    {
            printf("The coin is heads. You Win!");
    }
    else if ((tossResult == 0) && (userSelection == 'h'))
    {
            printf("The coin is tails. Better luck next time.");
    }
    else if ((tossResult == 1) && (userSelection == 't'))
    {
```

```
        printf("The coin is heads. Better luck next time.");
    }

}
```

The new output provides the user with a different output experience. Run the program a few times to test the code. The if-else-if ladder uses two conditions per rung to test the 4 possible outcomes. A final else-statement is not needed, since there can only be 4 results.

3.5 Switch-Case

Besides the if-else statement, another decision/branching statement is the switch-case statement. The switch-case statement is a control statement that evaluates an expression against a number of possible values and executes a specific block of code depending on the matching value. The switch-case statement is a compact implementation of the if-else-if ladder. It also provides for a default block of code that can be executed if none of the cases is satisfied. Notice that we used the word "expression" and not "condition". The following code listing is a simple example that checks an integer value.

```
#include <stdio.h>
#include <stdbool.h>

void main() {

        int planetnumber = 5;

        switch (planetnumber) {
        case 1:
            printf("Mercury");
                break;
        case 2:
                printf("Venus");
                break;
        case 3:
                printf("Earth");
                break;
        case 4:
            printf("Mars");
                break;
        case 5:
                printf("Jupiter");
                break;
```

```
        case 6:
                printf("Saturn");
            break;
        case 7:
            printf("Uranus");
                break;
        case 8:
                printf("Neptune");
            break;
        default:
                printf("Not a planet");
        }
    }
```

The switch statement checks the integer expression, "planetnumber", against the 8 available integer values. If it finds a match, the program executes what is in the block. If there is no match, the default case will be executed. You may think that this is similar to the if-else-if ladder and that the flowchart diagram would look very similar. The difference is that the switch-case is checked for a single expression, and the if-else-if ladder can have more complex conditions to test. For the switch-case, each case must match the same value type of the switch expression. The types supported by switch-case are String, char, int, byte, or short. You cannot mix types like strings and integers unless you do a caste or conversion ahead of time. The choice to use the if-else-if ladder versus the switch-case is dependent on the problem to be solved. For example, a group of three radio buttons in a graphical application may be assigned the values 0, 1, and 2. A switch-case can be used to determine the next process when the user selects one of the three radio buttons.

The break statement forces the program to jump out of the switch-case. If you forget the break statement, the program will continue to execute all the statements that follow. This is known as falling through. There might be situations where falling through can be exactly what you want. For example, several cases are to perform the same tasks. You can stack cases together and have a single block with a break statement. Here is a day-of-the-week example:

```
switch (day) {

        case 1: //Sunday
        case 7: //Saturday
                printf("The weekend\n");
                break;
```

```
case 2: //Monday
case 3: //Tuesday
case 4: //Wednesday
case 5: //Thursday
case 6: //Friday
        printf("Weekday\n");
        break;
}
```

3.6   Loops / Iteration

The if-statement and switch statement provided conditional branching for our programs. The next construct is iteration or loop statements that repeat the execution of code until a condition is reached.

Definition: *Iteration* is a repetition of a process to achieve a desired result.

Loops can programmatically solve basic math sequences like factorial (n!). From a practical perspective, loops can be used to calculate compound interest for a savings account or a dividend payout for a stock over a given time period. This section will introduce the different loop statements that are available in C.

3.6.1   While-Loop

The most basic loop is the while-loop. Here is the structure of the loop:

```
//while loop
while (condition)
{

}
```

The code is simply saying: "While the conditional expression is true, run the block of code, and continue to run the block of code until the conditional expression is false." Hence, the while-loop executes a group of statements enclosed in {} until a specified expression evaluates to false. The while-loop is similar to the if-statement, except the code block is repeated. The conditional test in the while-statement is at the beginning of the loop, so the loop block will never execute if the test initially fails. Any variables to be part of the conditional expression must be declared before the while-loop to be in scope at the time of the conditional test. The variables declared inside the while-loop will only be created if the loop is executed. The following code listing is an example:

```c
#include <stdio.h>

void main() {

    int x = 0;

    //While-loop test
    while (x < 10) //test if x is less than 10
    {
        printf("This is loop number %d\n", x);
        x++; // increment x on each pass through the loop
    }
    printf("x is now %d", x);
}
```

The variable x is an integer with an initial value assigned as 0. The while-loop tests to see if x is less than 10 and if the condition is true, then the code block is executed. The code block includes output to list the loop number, and the code block includes a statement to increment the value of x. The loop continues until x is no longer less than 10, which in this case is when x = 10. The value of x = 10 is tested in the while-loop and fails, the code block is not executed, and the program continues with the next statement after the while-loop. The final output displays the final value of x, which is now 10. A more complex condition with Boolean logic can be used with while-loops. The following flowchart is for this program.

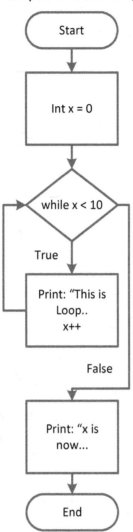

A common error with programming loops is never terminating the loop. If the program fails to generate a false expression, the loop can go on forever. We call this an infinite loop

Definition: *Infinite-loop* is a loop for which there is no terminating condition.

There are some microcontrollers that require an infinite loop at the end of a program. The following while loop is a purposeful infinite loop:

```
//Infinite loop
while (true)
{

}
```

We could end all the programs this way so we don't have to set breakpoints. Azure Sphere programs use an infinite loop to keep the system alive and running.

3.6.2   Do–While-Loop

The do-while-loop executes a group of statements enclosed in curly brackets "{}" repeatedly until a specified condition evaluates to false. The difference from the while-loop is that the statement or block of statements will execute at least one time before the expression is tested. Here is the basic structure:

```
//do-while loop
do {


} while (condition);
```

The test, in the do-while, is at the end of the loop, not at the beginning like the while-loop so the loop block will always execute at least once irrespective of the result of the test. The following code listing is an example.

```
#include <stdio.h>

void main() {

    //int x = 0;
    //
    ////While-loop test
    //while (x < 10) //test if x is less than 10
    //{
    //     printf("This is loop number %d\n", x);
    //     x++; // increment x on each pass through the loop
    //}
    //printf("x is now %d", x);

    int x = 10;
```

```
//do-while loop
do {
        printf("This is loop number %d\n", x);
        x++; // increment x on each pass through the loop

} while (x < 10);
printf("x is now %d", x);

}
```

The program looks very similar to the while-loop example, but this time x = 10. When the application runs, the code block is executed once and the loop is exited. Change x = 0, and run the program. Now, you will have the same output as the while-loop example. The flowchart for the do-while loop is as follows:

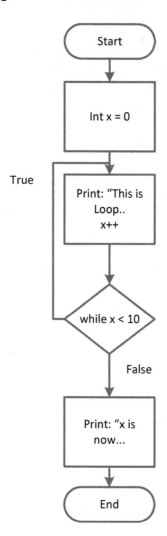

3.6.3   For-Loop

The most versatile loop is the for-loop. Here is the basic structure:

```
//For loop
for (initialization; condition; iteration) {

}
```

The for-loop executes a group of statements enclosed in {} repeatedly until a specified condition evaluates to false. The for-loop is an enhancement to the while-loop. The for-loop provides a mechanism for initializing variables before the loop begins and a mechanism for modifying variables after each pass through the loop. The test condition, like the while-loop, is performed at the beginning. The following code listing is the same program from the while-loop, but implemented with a for-loop:

```c
#include <stdio.h>

void main() {

        //For loop
        for (int x = 0; x < 10; x++) {
                printf("This is loop number %d\n", x);
        }

}
```

Here is the flowchart for the for-loop:

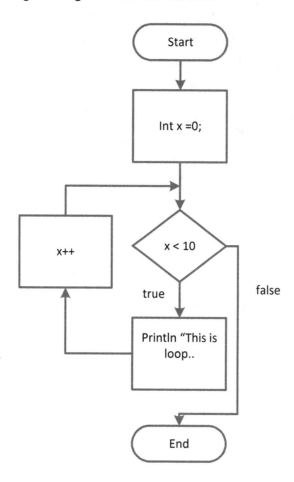

The output will be the same as the while-loop, with one exception. There is no last output statement as it is outside the for-loop scope. The scope of variables will be discussed at a later time.

There is also a particular variation of the for-loop that is expressed as follows:

```
for(;;);
```

This produces a tight, infinite loop. This is typically used as a way to halt a program without inserting a breakpoint and can halt the program without running it under the debugger. You will see examples of this in some of the computer activities later in the book. You can also use this in multi-tasking programs as a way to pause a thread while waiting for activity from another thread.

3.6.4   Computer Activity 3.4 – Factorial

This activity will involve a for-loop and an if-statement to solve an algebra problem. The factorial of a positive integer n is denoted by n! which is a product of all positive integers less than or equal to n. For example, 4! is 24 = 4 * 3 * 2 * 1. The factorial function is defined as:

$$n! = \prod_{k=1}^{n} k$$

Another way to present the equation n! = n * (n-1)!, thus 5! = 5 * 4!, 4! = 4 * 3!, etc. Of course 0! = 1. Create a new C project in Visual Studio called CH3-Factorial. Enter the following code:

```c
#include <stdio.h>
#include <string.h>

void main() {

        char userNumber[10];
        int fact = 1;


        printf("Calculate a factorial\n");
        printf("Enter an integer value for n\n\n");

        scanf_s("%9s", userNumber,10);

        int n = atoi(userNumber);

        if (!(n == 1 || n == 0)) {
                for (int x = 1; x <= n; x++) {
                        fact *= x;
                }
        }
        printf("The factorial for %d is %d\n", n, fact);

        while(1) {

        }
}
```

Run the program and enter different values of n but keep the value less than 10. You can use a calculator that supports "n!" to double-check to see if the calculations are correct.

The first part of the code asks the user for a value for n. The userNumber string is used to capture the user input. The value is then converted to an integer value. The fact integer variable is set to 1 since "1!" and "0!" equal 1. The if-statement checks for the case of 1 and 0. If the value is 1 or 0, the result is 1. If n doesn't equal 1 or 0, then the for-loop is executed to calculate the factorial. The variable x starts with 1 and is tested for the condition of less than or equal to n. The factorial value is simply the product of the variable fact multiplied by x. The product is calculated every iteration.

3.6.5   Break and Continue

Sometimes you want to jump out of the middle of a loop or just skip some processing in a loop to continue with the next iteration based on certain conditions. For example, a program might be searching through a group of records, and once it finds the record, it will stop searching. There are two statements that handle the situation. The first is the break-statement, which was covered in the switch-case statement. The break-statement simply breaks the code of execution and jumps out of the current block. The second is the continue-statement, which simply skips over any remaining statements to the next Boolean expression. The following code listing is an example of the break-statement in action.

```c
#include <stdio.h>

void main() {

        int x;
        for (x = 0; x <= 10; x++)
        {
                if (x == 6)
                {
                        break;
                }
                printf("this is loop number %d\n", x);
        }
        printf("the final number is %d", x);

        while (1) {

        }

}
```

Here is the output:

```
this is loop number 0
this is loop number 1
this is loop number 2
this is loop number 3
this is loop number 4
this is loop number 5
the final number is 6
```

When x equals 6, the program will break out of the loop. In this example, the loop never reaches the maximum value of 10. The following code listing is the same program, but with the continue-statement:

```c
#include <stdio.h>

void main() {

    int x;
    for (x = 0; x <= 10; x++)
    {
        if (x == 6)
        {
            continue;
        }
        printf("this is loop number %d\n", x);
    }
    printf("the final number is %d", x);

    while (1) {

    }

}
```

Here is the output:

```
this is loop number 0
this is loop number 1
this is loop number 2
this is loop number 3
this is loop number 4
this is loop number 5
this is loop number 7
```

```
this is loop number 8
this is loop number 9
this is loop number 10
the final number is 11
```

When x equals 6, the execution of the code moves on to the next iteration and never prints out the value of 6. The loop will continue until x equals 10.

3.6.6 Computer Activity 3.5 - Menu System with User Input

Combining a switch-case with a loop provides a structure to have a menu system in a command line application. The user can make selections based on a numbered item and then enter further information into the program. Create a new C project in Visual Studio called CH3-MenuSystem. Enter the following code:

```c
#include <stdio.h>
#include <stdbool.h>
#include <string.h>

void main() {

        bool closeprogram = false;
        char menuitem;
        char name[20];
        char address[50];

        do {

                printf("Enter the number to select the menu item\n");
                printf("1. Enter a name\n");
                printf("2. Enter an address\n");
                printf("3. Exit application\n");

                menuitem = getchar();
                getchar(); //dummy read to get the character in the buffer

                switch(menuitem)
                {

                    case '1':
                            printf("Enter a name:\n");
                            gets_s(name, 20);
                            printf("The name you entered is %s\n", name);
                            break;
```

```
                    case '2':
                            printf("Enter an address:\n");
                            gets_s(address, 50);
                            printf("The address you entered is: %s\n",
address);
                            break;

                    case '3':
                            closeprogram = true;
                            break;
                }
            printf("\n");

        } while (!closeprogram);
        printf("Exiting program\n");

}
```

Run the program and enter values for the menu selection to test the program. The application will continue to run until the user enters 3 to exit the application. Since getchar() only gets a single character, there is still the carriage return in the buffer so a second getchar() is called to clear the input buffer so that you can enter values for name and address without the application exiting prematurely.

3.7   Scope of Variables

The for-loop demonstrated that a variable defined within the for-loop is only available within the for-loop. This is known as the scope. *Scope* is the lifespan of a variable. Where it is created, where it is available to the rest of the code, and where it is destroyed are controlled by code blocks. The curly brackets {} (braces) define a code block. If a variable is defined within a code block, it is only available within that code block and any nested code blocks. The variable's scope is within the block. In the example below, the printf() outside of the for-loop will generate an error when attempting to compile the program. The variable x in the printf statement is undefined outside of the for-loop code block.

```
//For loop
for (int x = 0; x < 10; x++) {
    printf("This is loop number %d\n", x);
}
printf("The value of x is %d", x);
```

One could move the creation of x before the for-loop, and the code would compile and run.

```
//For loop
int x = 0;
for (; x < 10; x++) {
        printf("This is loop number %d\n", x);
}
printf("The value of x is %d", x);
```

Since programs are broken down into several library functions, you may want to limit the scope of variables. Variables defined with the reserved word static can only be accessed within the program the variable resides. Once created, they exist as long as the program is running. Their scope is limited to the block in which they are defined. They are not created on the stack.

```
static int y;
```

If a variable is going to be widely used in the program, a variable can be declared as register.

```
register int j;
```

A registered variable tells the compiler to put the variable into a CPU register. The reason to do this is to make the program a little faster. The limitation of using registered variables is based on the programming environment. Using the register type does not guarantee that the variable will actually be created using a CPU register. This is a suggestion and will default to a stack variable, if there are no CPU registers available.

Remember the variables and their scope when you get to Chapter 9 Memory Management. You will see where different variable types get placed into the overall memory map of an application.

3.8   Debugging

With the different branching, looping, and scope possibilities, programs can get more complex and error-prone. Errors are going to happen, but good coding practices and the right development tools will help to reduce the errors in code. There are three types of errors in a program: syntax errors, runtime errors, and logical errors. Syntax errors occur when you type something incorrectly in the editor. Most modern editors spot syntax errors fairly quickly. Runtime errors occur when the program is running and cannot perform a statement. If you started the program in Visual Studio, the program would stop with an exception indicated, if a runtime error occurs. Logic errors are the most difficult to find. A program may have the correct syntax and never have a runtime error, but once in a while, an unexpected result occurs. Logic errors require the right tools to track down the root cause of the error, and this is where Visual Studio's built-in debugger capability becomes helpful.

3.8.1   Computer Activity 3.6 – Debugging an Application

All the applications we have run so far have been run in debug mode. The real reason is that as we develop software, using the debugger is the best way to work out all the bugs before release. Let's take a simple look at the debugger. Create a new C project in Visual Studio called CH2-Debug. Enter the following code:

```c
#include <stdio.h>

void main() {

    for (int x = 0; x < 5; x++)
    {
        printf("The current index is %d\n", x);
    }

}
```

Now, we need to set a breakpoint in the for-loop at the printf call. Move the cursor to the left gray margin of the editor. As you move the mouse cursor up and down along the gray margin, a dark dot will appear next to the line that the mouse cursor is opposite. Move the mouse cursor next to the line for the printf statement and click the dark dot. The dark dot will change to red and stay at that location opposite the printf statement. This indicates that a breakpoint has been set for that line. If you want to remove the breakpoint, simply click on the red dot, and it will disappear. If you want to disable the breakpoint but leave it there so you can enable and use it later, right-click on the red dot, and from the drop-down click on "Disable Breakpoint". The red dot will change to a red circle indicating the breakpoint is there but inactive. Right-clicking the red circle, later, and selecting "Enable Breakpoint" will cause the breakpoint to be active, again, and the red circle will again become a red dot; or you can simply click on the red circle to activate the breakpoint.

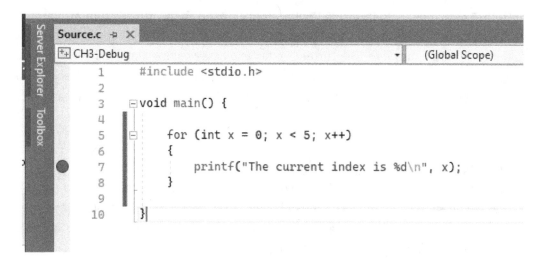

To debug the code, from the menu, select Debug->Start Debugging or hit F5. The application starts running and halts when it hits the breakpoint. Use the debug controls or use the corresponding function keys to step through the code. Click on the Step over or hit the F10 key a few times.

Function Key	Control	Debugger Action Taken
F11	↓	Step into. Jumps into a method or function if these are the next steps.
F10	↷	Step over. Jumps over a method or function. It will jump over and make these calls and move on to the next step.
Shift+F11	↑	Step out of a call.

As you iterate your way through the loop, you will see the x variable change values. The locals window shows the variables and their values. Stop debugging when finished.

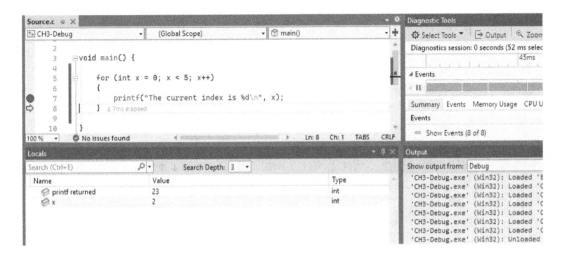

3.9 Summary

The chapter covered two very important topics: branching and iteration. Boolean expressions are used in both constructs to either branch the program into an alternate section of the code or break out of a loop. The program flow can get complex. The irony is that you almost have to think like the computer to avoid mistakes, which can be a challenge for large programs; but Visual Studio's debugger can help step through the code to see what is going on. We will build on these constructs and exercises throughout the rest of the book.

4 Functions and Program Structure

The Menu System project in the last chapter was one of the longest projects so far, which brings us to this chapter. As programs get bigger, they can get more complex to debug and maintain. The whole program existing in the main() function is not practical or recommended. Breaking down the project into smaller functions in the same or different files is the preferred coding method. In this chapter, we will explore creating different functions and general program structures.

4.1 Functions – Passing and Returning Data

The math functions discussed in Chapter 2 take in data and return a result. Some functions can simply be called to return a result like getchar(). How you design a function is up to you, but the function call and return methodology that must be followed in C is well defined. Let's dig into a computer activity to see how to create a function.

4.1.1 Computer Activity 4.1 – Temp Converter

We will create a function that converts a Fahrenheit value to a Celsius value. Create a new C project in Visual Studio called CH4-TempConvert. Enter the following code:

```
#include <stdio.h>

//Function Declaration.
double FtoC(double x);

void main() {

        double tempF = 80.1;
        //Call the function
        double tempC = FtoC(tempF);

        printf("The temperature %3.1lf Fahrenheit is %3.1lf Celsius\n",
 tempF, tempC);
```

```
}
```

```
//Implementation
double FtoC(double tempFahrenheit) {

        double celsius = ((tempFahrenheit - 32) * 5) / 9;

        return celsius;
}
```

Set a breakpoint on the printf() function in main() and a breakpoint at the Celsius calculation function. Run the application, and the program will stop at the breakpoint in the FtoC() function. Single step through the code until the result is displayed in the console window. The function FtoC() is first declared before main(), and the implementation of FtoC() comes after main(). This will become relevant when we talk about libraries a little later. The FtoC() function is declared as a double data type, which means the function is going to return a double value after being called. Following the function name, the round brackets encapsulate the arguments that the function will accept. For this example, "double x" is used for the declaration of the type of the argument to be passed into the function, but "double tempFahrenheit" is used for the declaration of the argument to be passed into the function in the actual function implementation. The declaration and the implementation don't have to have the same function argument names, but it is best practice to make the declaration of a function as descriptive as the implementation. Typically, the function declaration and implementation will match. Comment out the FtoC() function declaration and try to compile the program.

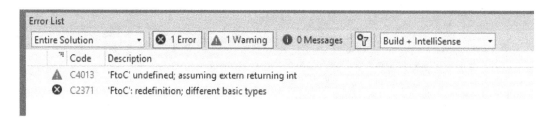

The C compiler cannot resolve the FtoC() function that main() is trying to use since it was not declared beforehand. Even if the function itself is in the same file, the compiler went from top down in the source.c file and threw the error when it couldn't resolve FtoC(). Declaring variables before use is the way the C language is defined. In other programming languages, like C#, function (method for C#) declaration is not needed. Now, move the FtoC() function implementation before the main() function and try to compile the project. The following code listing shows the change in the code.

```c
#include <stdio.h>

//Function Declaration
//double FtoC(double x);

double FtoC(double tempFahrenheit) {

        double celsius = ((tempFahrenheit - 32) * 5) / 9;

        return celsius;
}


void main() {

        double tempF = 80.1;
        //Call the function
        double tempC = FtoC(tempF);

        printf("The temperature %3.1lf Fahrenheit is %3.1lf Celsius\n",
tempF, tempC);

}

//Implementation
//double FtoC(double tempFahrenheit) {
//
//      double celsius = ((tempFahrenheit - 32) * 5) / 9;
//
//      return celsius;
//}
```

This time the compiler doesn't throw an error. Some developers will put main() at the very end of a program file. They will then implement all the functions before main() without having to avoid the extra code for a declaration.

4.2   Creating Separate Libraries

As putting all the code into a single main() is not practical, putting all your code into a single file is not a best practice either. You will want to re-use some of the functions that you create in other projects. A good way to do this is to create your own library in C. This will require a header file (.h) to provide the function declarations that get included into your project using the #include declaration.

4.2.1   Computer Activity 4.2 -Temp Converter 2

In the last computer activity, a simple Fahrenheit to Celsius function was created. In this project, we will create a separate source file that contains both a Fahrenheit to Celsius function and a Celsius to Fahrenheit function.

Create a new C project in Visual Studio called CH4-TempConvert2. There are a number of actions required, so we will break it down step-by-step.

1.   Once the project has been created, in the Solution Explorer on the right, right-click on the Source Files.
2.   Select Add->New Item... from the pop-up menu.
3.   An Add new Item dialog appears. You should see and be in Visual C++ context. Click on C ++ File (.cpp)
4.   In the Name section, enter tempconvert.c and click on the Add button.

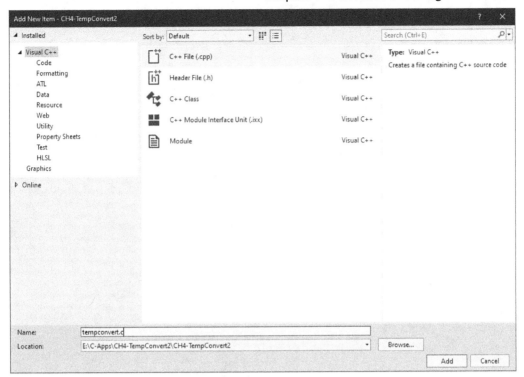

5. In Solution Explorer, right-click on Header Files.

6. Select Add->New Item… from the pop-up menu.

7. An Add new Item dialog appears. You should see and be in Visual C++ context. Click on Header File (.h).

8. In the Name section, enter tempconvert.h, and click on the Add button.

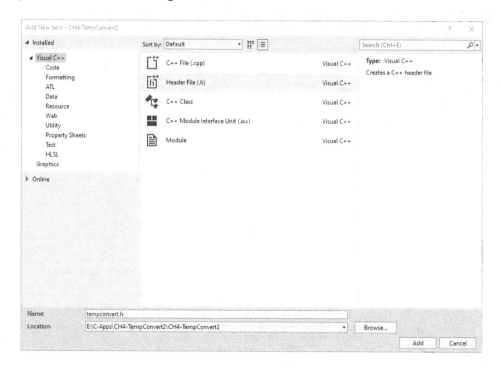

In Solution Explorer, you should see the two new files and the Source.c file.

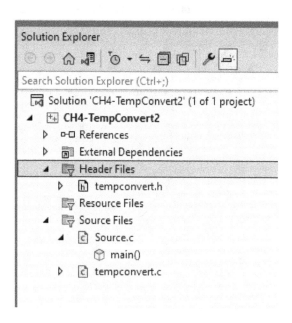

9. Open tempconvert.c, and add the following code:

```
#include "tempconvert.h"

//Function Implementations

double FtoC(double tempFahrenheit) {

        double celsius = ((tempFahrenheit - 32) * 5) / 9;

        return celsius;
}
double CtoF(double tempCelsius) {

        double fahrenheit = (tempCelsius * 9 / 5) + 32;

        return fahrenheit;
}
```

10. Save the file.
11. Open tempconvert.h. You will notice that a #pragma once is already in the file. The C Preprocessor discussion is coming next. Add the following code:

```
#pragma once

//Function declarations

double FtoC(double tempFahrenheit);

double CtoF(double tempCelsius);
```

The header file contains the function declarations. The header file will be included in the Source.c file so as to address the function declarations that main() can call.

12. Open Source.c, and add the following code:

```
#include <stdio.h>
#include "tempconvert.h"

void main() {

        double tempF = 75.5;
        double tempC = 23.2;
```

```
      printf("%3.1lf Fahrenheit is %3.1lf Celsius\n", tempF,
FtoC(tempF));
      printf("%3.1lf Celsius is %3.1lf Fahrenheit\n", tempC,
CtoF(tempC));
}
```

13. Set a breakpoint on the last printf() function in main(), and run the application.
14. The application will run, output the first message, and break on the second printf() call. Click step into.
15. The code jumps to the tempconvert.c and the CtoF() function. Step over the code until the program ends.

By putting like functions into a single file, you can create your own library. Debugging the source code in separate files is just like debugging code in a single file. Separating the code into different files makes it easier to debug and manage a big project. If you want the library to be in a standalone project so you can share the library with others, you would have to create a C library from a C++ library project template. As of this writing, we have not created a C library template.

4.3   The C Preprocessor

C provides C Preprocessor directives to facilitate compilation. We have been using #include in the projects to include the specific C libraries. *#pragma once* in the last computer activity is non-standard but is widely supported. *#pragma once* tells the compiler to only compile the code once. Here are the basic C Preprocessor directives:

- #include <header file name> - The compiler searches for the header file to include the declarations from the associated library. You may have to give the file a path if the file is not in one of the preprocessor's default search paths.

- #define name <replacement> – used to create a macro definition and define constants. You can create a token name that has a value or an operation to be performed.

- #if, #else, #elif, #endif, and #ifndef – Conditional compiler directives force the compiler to include something if a condition is met. A project might target different implementations.

4.3.1 Computer Activity 4.3 – C Preprocessor Examples

Let's see the preprocessor directives in action. Create a new C project in Visual Studio called CH4-Preprocessor. Enter the following code:

```c
#include <stdio.h>

#ifndef macroMath
#define macroMath
#endif // !macroMath

#define endLoopForever for(;;)

void main() {

    int x = 2;

#ifdef macroMath
#define sumX (x+x)
    printf("The sum of x + x = %d", sumX);
#else
    x *= x;
    printf("The square of x * x = %d", x);
#endif

    endLoopForever;
}
```

Run the program, and the program doesn't close right away. The endLoopForever macro runs an infinite for-loop so the program never terminates. The conditions of macroMath being defined or not, either produce the sum of the integer X or the square of X. Since macroMath is defined, the sum will be calculated and the square code is grayed out. Stop debugging the application. Comment out #define macroMath and re-run the program. The square of X is calculated. The sum code is grayed out and not included in the compilation.

4.3.2   Computer Activity 4.4 – Magic Numbers

If we look back at Computer Activity 3.5 - Menu System with User Input, we see the following char array definitions:

```
char name[20];
char address[50];
```

The values 20 and 50 are called Magic Numbers. There are a few definitions for Magic Numbers, but in this case, we are using a constant value as a literal. Proper coding would have been to use a constant name and assign it the value. This way, if the value is used in other locations, you can just use the symbolic name and you don't have to remember what the Magic Numbers were. If you use a Magic Number in several locations and then decide to change it, you have to be sure to change the Magic Number in every location that it is used. If you use a constant, you only have to change the one location where the constant is defined, and you won't have to worry about failing to update all of them correctly. In this computer activity, we will show the proper code technique that removes the magic numbers 20 and 50.

Create a new C project in Visual Studio called CH3-MagicNumbers. Enter the following code:

```c
#include <stdio.h>
#include <stdbool.h>
#include <string.h>

int main() {
#define NAME_ARRAY_LENGTH        20
#define ADDRESS_ARRAY_LENGTH     50

    bool closeprogram = false;
    char menuitem, dummy;

    char name[NAME_ARRAY_LENGTH];
    char address[ADDRESS_ARRAY_LENGTH];

    do {
        printf("Enter the number to select the menu item\n");
        printf("1. Enter a name\n");
        printf("2. Enter an address\n");
        printf("3. Exit application\n");
```

```
        menuitem = getchar();
        dummy = getchar(); //dummy read to get the character in
the buffer

        switch (menuitem)
        {
        case '1':
                printf("Enter a name:\n");
                gets_s(name, NAME_ARRAY_LENGTH);
                printf("The name you entered is %s\n", name);
                break;
        case '2':
                printf("Enter an address:\n");
                gets_s(address, ADDRESS_ARRAY_LENGTH);
                printf("The address you entered is: %s\n",
address);
                break;
        case '3':
                closeprogram = true;
                break;
        default:
                break;
        }

    } while (!closeprogram);
    printf("Exiting program\n");

}
```

If you look at this source code and compare it to the Computer Activity 3.5 - Menu System with User Input, you will see the first critical differences in the first 2 lines after int main():

```
#define NAME_ARRAY_LENGTH       20
#define ADDRESS_ARRAY_LENGTH    50
```

The magic numbers are replaced with constants using the #define compiler preprocessor directive. Every place each constant name is used in the source code the preprocessor with replace it with the #define'd value. You will see that in the definitions for the character arrays:

```
    char name[NAME_ARRAY_LENGTH];
    char address[ADDRESS_ARRAY_LENGTH];
```

You will also see it in the gets_s() get string calls:

```
gets_s(name, NAME_ARRAY_LENGTH);
```

And

```
gets_s(address, ADDRESS_ARRAY_LENGTH);
```

Now, if you choose to change the size of either or both of the arrays, you only need to make the change in the #define statements and all places where the constants are used in the code will be updated with the new value. This coding mechanism not only makes the source code more readable, it helps prevent coding errors when you are developing and modifying your programs.

4.4 Summary

Breaking a program down into different functions is an important part of creating well-maintained code. This chapter looked at how a function can take parameters and return a result. Creating custom libraries allows you to reuse code in other projects. Finally, the C preprocessor directives help structure what you want the compiler to include in the program based on the context of the code. We will provide more examples throughout the rest of the book.

5 Arrays and Pointers

When the topic of arrays and pointers is presented in C/C++ textbooks, pointers are presented first. Most students coming out of school are learning Java, Python, and C# and are not generally taught about pointers and pointer-like structures in basic programming courses for these languages. Since arrays will be more familiar to programmers familiar with higher-level programming languages like Jave, Python, and C#, we will implement a different pedagogical method by talking about arrays first and then pointers.

5.1 Arrays and the One-dimensional Arrays

An array is one of the first and simplest data structures (objects) used in computer programming.

Definition: *Array* (or Array Data Structure) is a collection of elements or variables of the same type. Each variable in the array is addressed by as many index values as the array has dimensions.

Arrays can be used to store data such as test scores, a list of daily temperatures, recipes, and addresses of contacts. A char[] array, a one-dimensional array of characters, was used to create a string. The power of using arrays is the ability to manipulate and operate on them. The different loops discussed in Chapter 3 can be used to access individual elements in an array and walk through any or all of the elements of the array under program control. The basic declaration format for one-dimensional arrays is as follows:

> Data Type <array name>[integer number of elements]

When an array is declared, the value for each element is initialized to zero. To assign values to an individual element of an array, you use the array name with an integer index, for example:

```
int scores[5];

scores[0] = 1;
```

```
scores[1] = 7;
scores[2] = 24;
scores[3] = 5;
scores[4] = 42;
```

The first index for an array is 0. The last index value is n-1, where n is the number of elements in the array. The first index for our "scores" array is 0, and the last index for our "scores" array is 4. The last index is not 5 since we start with 0. Many programmers have missed the 0 to n-1 indexing for arrays, so getting the length of an array is important, as well as, not trying to index an array beyond its defined size. Mentally picturing the contents of an array is important. You can think of each index as an address for the element, much like houses on a street. To visually represent an array, you could draw a box chart like Figure 5.1:

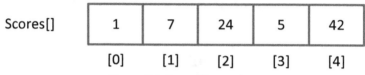

Figure 5.1 Array Diagram

The sizeof() function with a little mathematical manipulation can get the number of elements in an array:

```
int scoreslen = sizeof(scores)/sizeof(scores[0]);
```

As discussed in Chapter 2, each data type has a different byte size. The calculation gets all the bytes in the array divided by the byte size of one element in the array. In the example above, the total size of the array is 20 bytes and the size of an integer is 4 bytes. The result is 5 array elements.

For readability and simplicity, C also allows array values to be set in the declaration known as an array literal:

```
int scores[] = { 1,7,24,5,42};
```

Besides integers, you can also declare arrays of different types such as char[], double[], bool[], and float[].

5.1.1 Computer Activity 5.1 – Basic Arrays and the Power of the Debugger

With this activity, we will create a simple array, fill in a few values, print the contents of the whole array, and use the debugger to step through the array handling. Create a new C Project in Visual Studio and name the project CH5-BasicArray. Enter the following code:

```c
#include <stdio.h>

void main() {

        int x[10];

        x[0] = 14;
        x[1] = 17;
        x[2] = 21;
        x[3] = 3;
        x[4] = 10;

        int xlen = sizeof(x) / sizeof(x[0]);


        for (int y = 0; y < xlen; y++)
        {
                printf("Index %d holds the following value: %d\n",y, x[y]);
        }
}
```

The application fills in the first 5 elements and then prints all 10 elements in the array. Set a breakpoint at the printf() function and run the program. Click Continue to see the output for each index in the array. The values in indexes 5 through 9 hold a large negative value, which is just filled in by the compiler. Expand 'x' in the local variable pane to see all the elements that are in the array. The value of y changes on each iteration through the loop.

Name	Value	Type
Locals		▼ ₽ ×
Search (Ctrl+E)	Search Depth: 3 ▼	
▲ ⊘ x	0x0000006dfe53f8b8 {14, 17, 21, 3, 10, -858993460, -858993460, -858993460, -...}	int[10]
⊘ [0]	14	int
⊘ [1]	17	int
⊘ [2]	21	int
⊘ [3]	3	int
⊘ [4]	10	int
⊘ [5]	-858993460	int
⊘ [6]	-858993460	int
⊘ [7]	-858993460	int
⊘ [8]	-858993460	int
⊘ [9]	-858993460	int
⊘ xlen	10	int
⊘ y	3	int

5.1.2 Computer Activity 5.2 - Out-of-Bounds

Now, we are going to see what happens when we try to increase the index beyond the array length. The program will generate the average high temperature for one week of temperature readings. Create a new C Project in Visual Studio and name the project CH5-TemperatureArray. Enter the following code.

```c
#include <stdio.h>
#include <math.h>

void main() {

        double avgtemp = 0.0;
        double temperature[] = {71.6, 74.0, 75.1, 75.1, 73.7, 73.6, 72.0};

        for (int x = 0; x <= 7; x++) {
            avgtemp += temperature[x];
        }

        avgtemp = (double)(round((avgtemp /= 7) * 10)) / 10;
        printf("The average high temperature for the week is %3.1lf",
avgtemp);

}
```

After you save the project, a squiggly underline appears under temperature[x]. Visual Studio is already indicating that there is going to be a problem.

Run the project, and the result is an error in the calculation. Going out of bounds is a typical error when dealing with arrays. Using the length of the array elements is a best practice. In this case, simply changing the for-loop parameters to x < 7 solves the problem, and running the program produces the correct result.

5.1.3 Strings are char Arrays
Chapter 2 introduced strings as character arrays (char[]). Here is a simple example.

```
char str1[] = {"Welcome to programming in C."};

int str1len = sizeof(str1) / sizeof(str1[0]);

for (int x = str1len-1; x >= 0; x--) {

    printf("%c", str1[x]);
}
```

The string str1 is declared as a char array and initialized with a string. Each character in the string is a single array element. The for-loop prints each character in the array in reverse order. Here is the output:

```
.C ni gnimmargorp ot emocleW
```

Each element in the str1 array is accessed individually. The for-loop starts with the last element and decrements x on each iteration of the loop until x = 0. The <string.h> library contains a number of functions to manipulate strings, but the fact that strings are char arrays allows you to create your own custom functions.

5.2 *Two-dimensional Arrays*
So far, we have worked with one-dimensional arrays. C supports two-, three-, or up to whatever the compiler supports. Multidimensional arrays are known as arrays of arrays. A two-dimensional array is the simplest form of a multidimensional array. A two-dimensional array is declared as follows:

Data Type <array name>[integer number of elements][integer number of elements]

You can initialize the size of the multidimensional array:

```
int x[5][8];
```

You can also declare an array with a literal using prepopulating data:

```
        int  scores[4][5]  =  {{1,6,7,8,9},  {5,10,4,13,6},  {6,8,2,1,4},
{23,4,1,7,17}};
```

Each grouping of numbers represents a row. The best way to visualize the two-dimensional array, scores[x][y], is to use a chart.

Scores[][]	[][y = 0]	[][y = 1]	[][y = 2]	[][y = 3]	[][y = 4]
[x = 0][]	1	6	7	8	9
[x = 1][]	5	10	4	13	6
[x = 2][]	6	8	2	1	4
[x = 3][]	23	4	1	7	17

You can think of the two indices as the coordinates of a spreadsheet or the address of a city block. One example of using a two-dimensional array is to store information like student test scores. Each student is assigned a row and each column is a test. All the elements or cells will store the score for the test. Another example: a farmer wants to map crop production based on a land grid map. Each cell in the two-dimensional array represents an acre of land. Finally, think of all the games that use a grid: tic-tac-toe, reversi, chess, checkers, Battleship®, Connect Four™, and Minesweeper just to name a few. Each cell is used to store a game piece location.

5.2.1   Computer Activity 5.3 – Two-dimensional Array
For this activity, we will cover two concepts in one program. The first test will be the basic declaration and how to store the data in a two-dimensional array. We will also see how to get the length of rows and columns, which will allow us to do a search on a two-dimensional array. Create a new C project in Visual Studio and name the project CH5-2DArray. Enter the following code:

```
#include <stdio.h>

void main() {

        int test1[4][5];
        test1[0][0] = 5;
        test1[1][3] = 3;
        printf("The value at 1,3 is %d\n", test1[1][3]);
```

```
        int rowlen = sizeof(test1) / sizeof(test1[0]);
        int collen = sizeof(test1[0]) / sizeof(test1[0][0]);
        printf("There are %d elements in the row\n", rowlen);
        printf("There are %d elements in the column\n", collen);

        int count = 0;
        int test2[3][4] = {{3,2,6,5},{2,7,8,1},{4,2,7,2}};
        int rowlen2 = sizeof(test2) / sizeof(test2[0]);
        int collen2 = sizeof(test2[0]) / sizeof(test2[0][0]);
        for (int x = 0; x < rowlen2; x++)
        {

            for (int y = 0; y < collen2; y++)
            {
                    if (test2[x][y] == 2)
                    {
                            count++;
                    }
            }
        }
        printf("The number 2 was found %d times\n", count);

        for (;;);

}
```

Note the use of the for (; ;); infinite loop to pause the program as discussed in Section 3.6.3.

Set a breakpoint at the test1 declaration and start the debugger. Step over the code until you get to the first for-loop. If you expand the multidimensional arrays in the Locals window, you can see the contents of the arrays. The compiler filled in a big negative number for each element until a value is assigned.

Locals		▾ ♯ ✕
Search (Ctrl+E) 🔍 ▾ ↑ ↓ Search Depth: 3 ▾		
Name	Value	Type
collen	5	int
collen2	4	int
count	0	int
rowlen	4	int
rowlen2	3	int
▲ test1	0x0000002a0f5dfb70 {0x0000002a0f5dfb70 {5, -858993460, -858993460, -8589...	int[4][5]
▲ [0]	0x0000002a0f5dfb70 {5, -858993460, -858993460, -858993460, -858993460}	int[5]
[0]	5	int
[1]	-858993460	int
[2]	-858993460	int
[3]	-858993460	int
[4]	-858993460	int
▲ [1]	0x0000002a0f5dfb84 {-858993460, -858993460, -858993460, 3, -858993460}	int[5]
[0]	-858993460	int
[1]	-858993460	int
[2]	-858993460	int
[3]	3	int
[4]	-858993460	int
▷ [2]	0x0000002a0f5dfb98 {-858993460, -858993460, -858993460, -858993460, -858...	int[5]
▷ [3]	0x0000002a0f5dfbac {-858993460, -858993460, -858993460, -858993460, -858...	int[5]
▲ test2	0x0000002a0f5dfc38 {0x0000002a0f5dfc38 {3, 2, 6, 5}, 0x0000002a0f5dfc48 {2,...	int[3][4]
▲ [0]	0x0000002a0f5dfc38 {3, 2, 6, 5}	int[4]
[0]	3	int
[1]	2	int
[2]	6	int
[3]	5	int
▷ [1]	0x0000002a0f5dfc48 {2, 7, 8, 1}	int[4]
▷ [2]	0x0000002a0f5dfc58 {4, 2, 7, 2}	int[4]
x	-858993460	int

5.3 Pointers

Why do pointers exist in C? All programming languages are abstractions of the hardware that they are running on. The C programming language is a close abstraction of the hardware. It does not hide the underlying hardware operations. Instead, it provides a generalized set of features for operations all hardware have in common. This feature-set is easily ported to any hardware and gives the programmer nearly all the capabilities of the assembly language without having to learn the assembly language for each and every hardware platform that C has been ported to. One of the common features of nearly all hardware platforms is the way that data is stored and accessed. Data, whether it be operating instructions or the data read from and written to is stored in a memory device at a particular address in that memory. The address of memory storage is the pointer. Most higher-level programming languages abstract the memory access so, at a programming level, you are insulated from the storage and recovery mechanism of the hardware. C does not. C gives you access not only to the value of a variable but to its address as well. When we declare int x or char str[], the names of the variable and the array are friendly names for the memory data and address where the data is stored. When we use the & operator, as in &x, we get the address of x and can store that in a pointer variable, say int* iPtr = &x. Then we can use the pointer to read from or write to that address, as in *iPtr = 5. These two code lines of code:

```
int y = 4;
printf("y has the value of %d at address %p\n", y, &y);
```

would result in something similar to:

y has the value of 4 at address 000000FAC92FF8D4

The variable y contains the value 4, and the memory address of y is displayed by using &y. %p is used in the printf() function to output the address in hexadecimal format. The address value would be different each time the program runs, which is more of a function of the operating system kernel than the program.

A pointer is a variable that contains the memory address of a variable. The indirection * symbol is used to declare pointers. For example:

datatype* ptr

A pointer has to be defined with a data type. The pointer points to a memory location that contains a specific data type. Building on the previous example, the code sample below shows a pointer yptr being declared. The pointer's address is assigned the address of the variable y. yptr points to a memory location with the value of 4. When the printf() function is called, the result would be 4.

```
int y = 4;
printf("y has the value of %d at address %p\n", y, &y);
int *yptr;
yptr = &y;
printf("%d\n", *yptr);
```

The big question is: "Why do we need pointers?". Pointers are sometimes the only solution to perform certain operations that will be demonstrated in future exercises, and using pointers often results in more compact and efficient code. Pointers are critical to giving functions not only access to data but also to the address of data when calling a function. The benefit of this will also be described later.

Arrays and pointers have a symbiotic relationship. Since there is a bit of a stigma to pointers, we specifically discussed arrays first to help better understand pointers so you will understand their power and usefulness. Hands-on exercises provide a better experience.

5.3.1   Computer Activity 5.4 – Pointers, Variables, and Addresses

The following program will create a couple of variables and a pointer that points to one of the variables. The values and memory locations will be printed out. A change will be made and we will see what actually gets changed. Create a new C project in Visual Studio and name the project CH5-BasicPointers. Enter the following code:

```c
#include <stdio.h>

void main() {
        int x = 5;
        int* ptr;
        int y;

        ptr = &x;
        y = *ptr;

        printf("X is %d, &x is %p\n", x, &x);
        printf("*ptr is %d, ptr is %p, and &ptr is %p\n", *ptr, ptr,
&ptr);
        printf("Y is %d\n", y);

        *ptr = 9;

        printf("X is %d, &x is %p\n", x, &x);
        printf("*ptr is %d, ptr is %p, and &ptr is %p\n", *ptr, ptr,
&ptr);
        printf("Y is %d\n", y);

        for (;;);
}
```

Run the program and you will get something similar to the following:

```
X is 5, &x is 000000A18256F5E4
*ptr is 5, ptr is 000000A18256F5E4, and &ptr is 000000A18256F608
Y is 5
X is 9, &x is 000000A18256F5E4
*ptr is 9, ptr is 000000A18256F5E4, and &ptr is 000000A18256F608
Y is 5
```

The pointer, ptr, points to the memory location X, which contains the value of 5. The picture below provides a computer memory representation of ptr and X. ptr has the address of X.

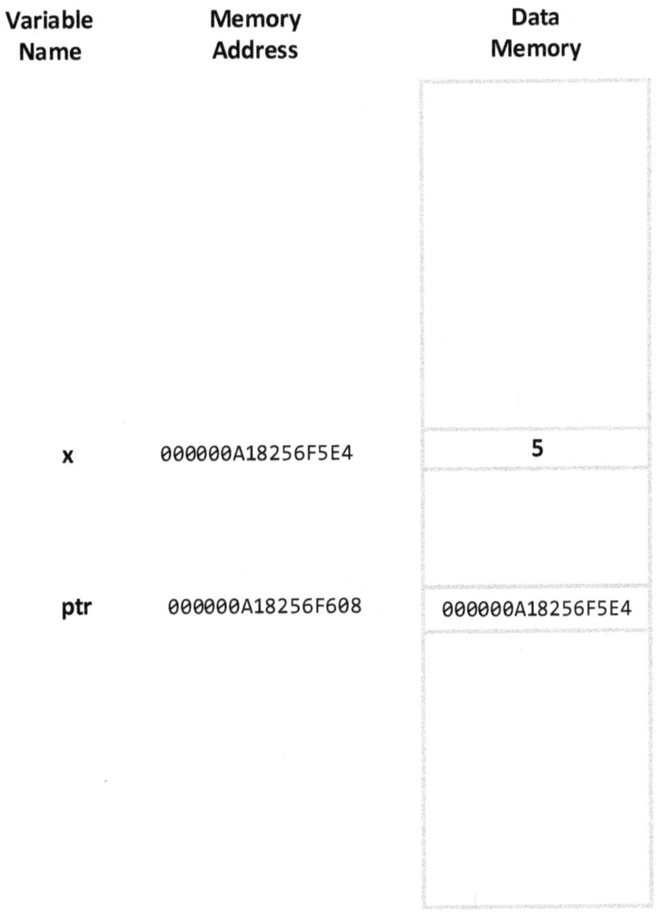

Variable Name	Memory Address	Data Memory
X	000000A18256F5E4	5
ptr	000000A18256F608	000000A18256F5E4

The Y variable is assigned the value of 5 via the ptr. Then the *ptr is assigned the value of 9, but since this is the indirect memory location of X, the value of X is now 9. The value of Y never changes.

Variable Name	Memory Address	Data Memory
x	000000A18256F5E4	9
ptr	000000A18256F608	000000A18256F5E4

5.3.2 Computer Activity 5.5 – Swapping Variables

The last exercise demonstrated a basic concept of indirection and manipulating variables with a pointer. In this program, we will use pointers to swap two variables. Create a new C project in Visual Studio and name the project CH5-SwapVariables. Enter the following code:

```
#include <stdio.h>

void swapvaraibles (int* variable1, int* variable2)
{
        int temp = *variable1;
```

```
        *variable1 = *variable2;
        *variable2 = temp;
}


void main() {

        int a = 6;
        int b = 1;

        printf("a = %d\n", a);
        printf("b = %d\n", b);

        swapvaraibles (&a, &b); //send the address of a and b to pointers

        printf("a = %d\n", a);
        printf("b = %d\n", b);

        for (;;);
}
```

Run the program and you will get the following:

```
        a = 6
        b = 1
        a = 1
        b = 6
```

The swapvariables() function swaps the values of two variables, but since a function can only return one value, pointers are used to manipulate both variables. The addresses of the variables are passed to pointers in the function. The pointers perform the swap of the values with the aid of a temp variable. Since the pointers are indirectly pointing to the locations of a and b, the function doesn't have to return anything. The function demonstrates a method that can only be solved with pointers.

5.4   Pointers and Arrays

The past exercises have shown how a pointer can point to a single variable location, and how it can manipulate the variable values without having to use the variable by name. The power of pointers becomes a little more interesting when it comes to arrays. When it comes to arrays, single- or multi-dimensional arrays, a single pointer can point to any element in an array. Let's start with a basic example.

5.4.1 Computer Activity 5.6 – Basic Pointer and Array

Create a new C project in Visual Studio and name the project CH5-BasicArrayPointer. Enter the following code:

```c
#include <stdio.h>

void main() {

        int scores[] = { 4,6,9,10,2,1 };

        int* scoreptr = scores;

        printf("scores address is %p\n", scores);
        printf("scoreptr is pointing to score address %p\n", scoreptr);
        printf("scoreptr is currently pointing to the value %d\n",
*scoreptr);

        scoreptr++;

        printf("scores[1] address is %p\n", &scores[1]);
        printf("scoreptr is pointing to score address %p\n", scoreptr);
        printf("scoreptr is currently pointing to the value %d\n",
*scoreptr);

        scoreptr += 2;

        printf("scores[3] address is %p\n", &scores[3]);
        printf("scoreptr is pointing to score address %p\n", scoreptr);
        printf("scoreptr is currently pointing to the value %d\n",
*scoreptr);

        for (;;);
}
```

Run the program, and you will get output similar as follows:

```
scores address is 0000008B180FF938
scoreptr is pointing to score address 0000008B180FF938
scoreptr is currently pointing to the value 4
scores[1] address is 0000008B180FF93C
scoreptr is pointing to score address 0000008B180FF93C
scoreptr is currently pointing to the value 6
scores[3] address is 0000008B180FF944
scoreptr is pointing to score address 0000008B180FF944
scoreptr is currently pointing to the value 10
```

The program declares and initializes the scores[] array. A pointer is then declared that points to the scores array and is actually pointing to the first element in the array, scores[0]. The debugger shows the value and address the pointer is pointing to.

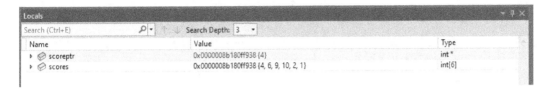

To get the address of the array, you just need "scores" without the brackets []. The & is not needed. The printf() statements that immediately follow, show the address location for score[0], the address the pointer is pointing to, score[0], and the value the pointer is pointing to.

After the printf() functions are called, scoreptr++ is then called. The address that the pointer is pointing to has incremented, which is actually 4 bytes since this is a pointer for integers and integers are 4 bytes long. The pointer is now pointing to scores[1]. The next set of printf() functions show the change. Finally, the address the pointer is pointing to is increased by 2 to point to scores[3]. The ability to manipulate what the pointer points to is the real power of using pointers.

5.5 *Pointer Arithmetic and Manipulation*

A single variable, pointer, with the ability to address multiple variables by changing the address allows you to create faster and more efficient code that takes up less memory. Arrays are easy to learn, which is why they were covered first in this chapter. Pointer arithmetic and manipulation is what some have found to be a challenge. We have specifically developed the first three-pointer computer activities to show the address of the pointer and the variable to demonstrate indirection. The last computer activity demonstrated the arithmetic to change the pointer's address value to point to a different location in an array. Pointers can be incremented, decremented, or offset. Pointers can point to specific memory locations, which will be covered in a later chapter. Relational operators such as <, >, >=, !=, etc. can be used to compare pointers.

5.5.1   Computer Activity 5.7 – String Reverse
In this computer activity, we will implement the string reverse program that we demonstrated earlier. A pointer will be used to reverse the string. Create a new C project in Visual Studio and name the project CH5-StringReverse. Enter the following code:

```
#include <stdio.h>

void main() {

    char str1[] = { "Welcome to programming in C." };

    int strlength = (sizeof(str1) / sizeof(str1[0]))-1;
    char* pstr1 = str1;
    pstr1 += strlength; // point to the end of the string

    for (; pstr1 >= str1; pstr1--) {
        printf("%c", *pstr1);
    }

    for (;;);
}
```

Run the program and the string will be printed in reverse. The array is declared and initialized with the characters. The pointer is declared and points to the first element in the array. Before the for-loop, the pointer is offset to the last element in the array. The for-loop uses the addresses for pstr1 and str1 for the conditional comparison. There is no need to define an extra integer. The pointer decrement is used to print each letter in reverse order.

5.5.2 Computer Activity 5.8 – String Reverse with No Array
Do we really need the array? Create a new C project in Visual Studio and name the project CH5-StringReverse2. Enter the following code:

```
#include <stdio.h>
#include <string.h>

void main() {

    char* str2 = { "Welcome to programming in C." };
    char* temp = str2;         //create a temporary pointer to store
the starting address
    str2 += strlen(str2); //set pointer to the end of the string

    while ( str2 >= temp) {

        printf("%c", *str2);
        str2--;
    }

    for (;;);
```

```
}
```

Run the program and the string is reversed again. As you can see, we don't need an array. The str2 pointer is declared and initialized with a string directly. A second pointer points to the beginning of the string, and the str2 pointer is then set to point to the end of the string using strlen() function from the <string.h> library. The pointer address is used as the condition in the while-loop, and the pointer decrement walks the pointer back through the string.

5.5.3   Computer Activity 5.9 – Alternative String Reverse with No Array

There is another way to do the string reversal without using the array, and letting a for-loop do all the heavy lifting. All the pointer manipulation is maintained in the scope of the for-loop and the original string and string pointer are never modified. Create a new C project in Visual Studio and name the project CH5-StringReverse3. Enter the following code:

```c
#include <stdio.h>
#include <string.h>

int main() {
        //Use a pointer to the string to be reversed, not an array
        char* strPtr = { "Welcome to programming in C." };

        //All reversing happens in the for-loop without changing or
duplicating the original string
        //The scope of the offset variable, i, is restricted to the
for-loop
        for (int i = strlen(strPtr); i >= 0; i -= sizeof(char))
        {
                //Print the string without modifying the string pointer
                printf("%c", *(strPtr + i));
        }

        //Pause forever
        for (;;);

}
```

Run this version of the program; and again, you will see the string printed in reverse. Look through the source code and pay particular attention to the comments. They point out all the key features of this approach. You can see that the for-loop, indeed, does all the heavy lifting; and by using the construct of the for-loop, it nicely documents what we are doing at a glance. The for-loop also keeps the offset variable nicely protected inside the scope of the for-loop, and the way we chose to use the pointer to select the character to be printed in each iteration of the loop, never modifies the string pointer itself. These concepts can be subtle when new to programming, but think about them and add them to your bag of tricks going forward.

5.5.4   Computer Activity 5.10 – Find the Max Value in an Array

The program will create an integer array of 10 numbers and then call a function to find the maximum integer value in the array. Create a new C project in Visual Studio and name the project CH5-FindMaxValue. Enter the following code:

```c
#include <stdio.h>
#include <time.h>
#include <stdlib.h>


int maxvalue(int* arrayptr) {

        int* scan = arrayptr;
        int* maxv = arrayptr;

        while (scan - arrayptr < 10) {
                if (*scan > *maxv) {
                        maxv = scan;
                }
                scan++;
        }
        return(*maxv);
}


void main() {

        srand((unsigned int)time(NULL));
        int x[10];

        printf("x array = ");

        for (int i = 0; i < 10; i++) {
                x[i] = (int)rand() % 100;
                printf("%d, ", x[i]);
        }
```

```
    printf("\n");

    int y = maxvalue(x);

    printf("Max value found is %d", y);


    for (;;);
}
```

Run the program a few times to see if the arrays get randomly filled and if the program finds the max value each time.

```
x array = 82, 88, 57, 46, 15, 7, 61, 64, 7, 58,
Max value found is 88
```

Now that we have pointers to work with, it's time to look at what pointers bring to the party with respect to calling functions and passing data to a function. The two major methods for passing data to a function are known as pass-by-value and pass-by-reference. Pass by value is what we have been doing up until now, and is exactly what it says. You pass the value of one or more variables to the function. Implicit in this is that a copy of the value is made by the function for its use. An equivalent variable is created by the function and the value of the variable being passed is copied into it. The variable whose value is being passed is untouched by the function. Any changes that the function may make to the variable that has the copy are only made to the function's copy, not to the variable being passed.

Now, let's look at pass-by-reference. In general, a reference to a variable is a means to access that variable and not only read its value but also write to that variable and change its value. In C, the reference mechanism is the pointer. If I have a pointer to a variable, I can read and write that variable. So, if I call a function and do a pass-by-reference of a variable, a copy of the pointer to that variable is sent to the function. This mechanism is both powerful and potentially dangerous. It's powerful because it allows you to give access to any size of variable using only a pointer. I don't have to make a copy of the whole variable. It's also powerful because it gives a method to send back multiple variables from the function instead of being restricted to the one-variable return. It is dangerous because I can change the value of the variable. The value of the variable after the function call may be different from the value of the variable after the function call. You have to be sure that any changes to variables passed-by-reference are what was intended by design, otherwise, you could be in for a lot of work debugging a bug introduced by a misuse of a passed pointer.

Now, look closely at the program. It fills an array with random numbers from 0 to 99. The array is passed-by-reference by passing the address, in a pointer, to the maxvalue() function. Two pointers are created that point to the array. One pointer, scan, is used to walk through every member of the array, and the other, maxv, stores the address of the current largest array value. As the scan walks the array and points to each value, the if-statement compares the value scanned and maxv value pointed to. If the scanned value is the larger than the maxv value, maxv is updated with the larger value's address. Once all the elements in the array have been scanned, maxv will point to the maximum value in the array; and that value will be returned. Where there might be confusion is the condition in the while-loop. Specifically, what does "scan – arrayptr" resolve to since these are address values. Add the following line between the if-statement and the scan++:

```
printf("value %lld\n", (scan - arrayptr));
```

Run the program again, and you can see the resulting value goes from 0 to 9.

```
x array = 39, 67, 48, 4, 61, 65, 42, 21, 60, 15,
value 0
value 1
value 2
value 3
value 4
value 5
value 6
value 7
value 8
value 9
Max value found is 67
```

5.5.5   Computer Activity 5.11 – Pointers to Multidimensional Arrays
Now let's look at how pointers interact with Multidimensional Arrays. Multidimensional arrays are simply arrays of arrays. Multidimensional arrays and each element in a multidimensional array can be accessed via its array index or via a pointer. The pointer to, address of, a multidimensional array is the address of the first element in that array. There can be arrays of nearly any data type, as well as, arrays of pointers.

Create a new C project in Visual Studio and name the project CH5-Pointer2DArrays. Enter the following code:

```c
#include <stdio.h>

void main() {

    int x[3][4] = { {1,2,3,4},{8,7,6,5},{0,9,3,2} };
    char* str1[] = { "Pointers", "are", "fun","!" };


    //Integer 2D array
    int* ptr;
    int(*ptrarray)[4];

    ptr = x;
    ptrarray = x;

    printf("Address pointed to by ptr = %p\n", ptr);
    printf("Current valued for ptr = %d\n", *ptr);
    printf("\n");
    printf("Address pointed to by ptrarray[] = %p\n", ptrarray);
    printf("Address pointed to by ptrarray[0] = %p\n", ptrarray[0]);
    printf("Current valued for ptrarray[0] = %d\n", (*ptrarray)[0]);
    printf("Address pointed to by ptrarray[1] = %p\n", ptrarray[1]);
    printf("Current valued for ptrarray[1] = %d\n", (*ptrarray)[1]);
    printf("Address pointed to by ptrarray[2] = %p\n", ptrarray[2]);
    printf("Current valued for ptrarray[2] = %d\n", (*ptrarray)[2]);
    printf("Address pointed to by ptrarray[3] = %p\n", ptrarray[3]);
    printf("Current valued for ptrarray[3] = %d\n", (*ptrarray)[3]);

    printf("\n");
    ptr++;;
    ptrarray++;

    printf("Address pointed to by ptr = %p\n", ptr);
    printf("Current valued for ptr = %d\n", *ptr);
    printf("\n");

    printf("Address pointed to by ptrarray[] = %p\n", ptrarray);
    printf("Address pointed to by ptrarray[0] = %p\n", ptrarray[0]);
    printf("Current valued for ptrarray[0] = %d\n", (*ptrarray)[0]);
    printf("Address pointed to by ptrarray[1] = %p\n", ptrarray[1]);
    printf("Current valued for ptrarray[1] = %d\n", (*ptrarray)[1]);
    printf("Address pointed to by ptrarray[2] = %p\n", ptrarray[2]);
    printf("Current valued for ptrarray[2] = %d\n", (*ptrarray)[2]);
    printf("Address pointed to by ptrarray[3] = %p\n", ptrarray[3]);
    printf("Current valued for ptrarray[3] = %d\n", (*ptrarray)[3]);


    //String 2D Array
    printf("%s %s %s%s \n", str1[0], str1[1], str1[2], str1[3]);
```

```
        for (;;);
}
```

Set a breakpoint at the first printf() function and run the program. If you look at the Locals pane, you will see the pointers to x point to different items. The ptr pointer points to the first element in the 2D array. The array of pointers, ptrarray[], points to the first row in the x 2D array.

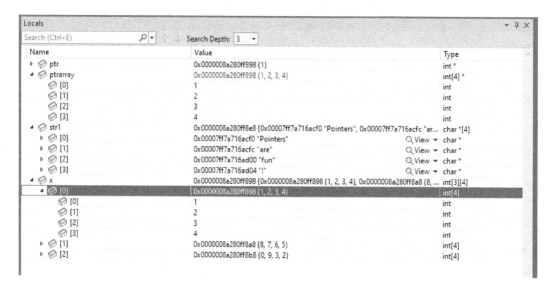

Step through the code and after the increment, ptr points to the next element, but ptrarray is pointing to the next row.

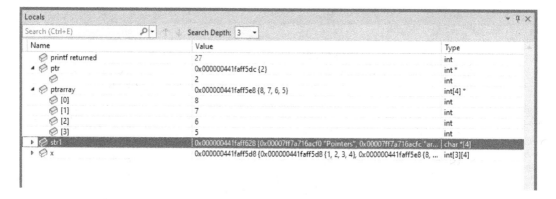

The x 2D array is thought of in terms of columns and rows, but in the physical memory storage, the elements are addressed linearly in memory. A 3x4 array is 12 total elements.

If ptr was incremented each time in a for-loop from 0 to 11, it would point to each element in the 2D array in succession. The ptrarray pointer defines an array of 4 pointers to the first 4 elements in the 2D array. When incremented, ptrarray moves to the next 4 elements. The pointer to the array of strings, str1[], follows a similar logic, where each string has a pointer.

5.5.6   Computer Activity 5.12 - Pointer Efficiency

Pointers make for more efficient code. Sometimes you might code something, but the compiler may also help to find a more efficient implementation. Let's take a look at a program. Create a new C project in Visual Studio and name the project CH5-PointerEfficiency. Enter the following code:

```c
#include <stdio.h>


void DoSomething(int localarray[], int size) {

        for (int x = 0; x < size; x++) {

                localarray[x] += (50 - x);
        }
}


void main() {

        int testarray[] = { 30,26,18,3,7 };

        DoSomething(testarray, 5);

        for (int x = 0; x < 5; x++) {
                printf("%d ", testarray[x]);
        }
        printf("\n");

        for (;;);
}
```

Set a breakpoint at the for-loop in the DoSomething() function. Run the program. Step through the program. Looking at the Locals window, you will notice that the localarray[] is not listed as an array, but as a pointer to the original array.

Name	Value	Type
◢ ⬦ localarray	0x00000078b5b1fc18 {80}	int *
⬦	80	int
⬦ size	5	int
⬦ x	0	int

Locals — Search (Ctrl+E) — Search Depth: 3

When DoSomething() function was called the address to the array was passed to the function. The compiler didn't open memory space to create a whole new array, but instead created a pointer behind the scenes. When you step out of the DoSomething() code, and back to main(), you will see the testarray[] and all the elements in the Locals window.

Name	Value	Type
◢ ⬦ testarray	0x00000078b5b1fc18 {80, 75, 66, 50, 53}	int[5]
⬦ [0]	80	int
⬦ [1]	75	int
⬦ [2]	66	int
⬦ [3]	50	int
⬦ [4]	53	int
⬦ x	-858993460	int

Locals — Search (Ctrl+E) — Search Depth: 3

Here is an equivalent implementation for DoSomething() that takes in a pointer as a parameter.

```
void DoSomething2(int* localarray, int size) {

    for (int x = 0; x < size; x++) {
        *localarray = (50 - x) + *localarray++ ;
    }
}
```

5.6   Command Line Arguments

You are probably familiar with command line utilities and the arguments (options) that can be passed into them. For example, you can query the services in Windows:

```
sc.exe query
```

All the computer activities that we have performed are command line utilities. The programs can also take in arguments at the command line. To do this, we add to the main() function two parameters: an integer and an array of character pointers.

```
main(int argc, char *argv[])
```

The integer contains the count of all items entered on the command line including the program and the arguments passed into the program. The array of character pointers points to each string argument that you have entered.

5.6.1   Computer Activity 5.13 –Basic Command Line Arguments

Create a new C project in Visual Studio and name the project CH5-Arguments. Enter the following code:

```c
#include <stdio.h>
#include <string.h>

int main(int argc, char *argv[]) {

    printf("Command line argumnets\n");

    if (argc <2 ) {
        printf("Usage: CH5-Argument -a <text> | -b | -c \n");
        return -1;
    }
    else if (!strcmp(argv[1],"-a")) {
        printf("You entered option -a and the text is: %s\n",
argv[2]);
    }
    else if (!strcmp(argv[1], "-b")) {
        printf("You entered option -b\n");
    }
    else if (!strcmp(argv[1], "-c")) {
        printf("You entered option -b\n");
    }
    else {
        printf("No option selected\n");
        printf("Usage: CH5-Argument -a <text> | -b | -c \n");
        return -1;
    }
}
```

This is a change from previous programs, the main() function to return an integer value, as well as, accept command line arguments. To debug with arguments, a test argument can be put into the build.

1.   From the menu, select Project-> CH5-Arguments Properties

2. In the Properties Pages dialog, under Configuration Properties->Debugging->Command Arguments, enter the following:

    -a "Hello World"

3. Click Ok.

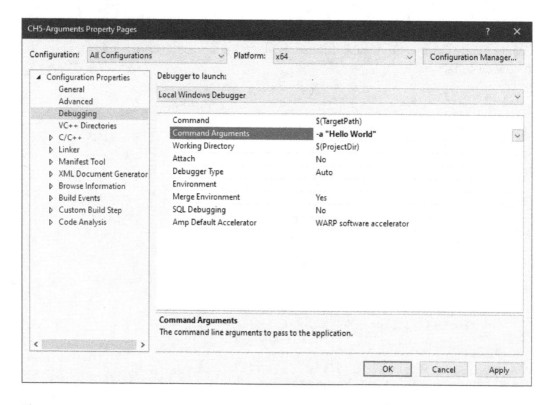

The quotations around Hello World tells the program to treat both words as a single command line argument. Without the quotes, Hello and World would be separate arguments because of the space character between them. Run the program and you will get the output for option -a and the string will be printed. You can also run the program from the command line:

    CH5-Arguments -b

The program gets the count from the command line and passes it in argc. The start address for each string in the command line is put into a separate element in the argv[] pointer. If you only entered the program name, the first if statement outputs a usage message and terminates the program. The rest of the if-else ladder performs a string comparison on the second item on the command line, which is the first argument. If there is a match, the corresponding printf() is called. If a wrong option is entered, the usage message is printed. Add the following after the if-else ladder, and re-run the program to get the output for the addresses and the starting location of each string as well as the string contents.

```
printf("Pointer Location %p\n", &argv[0]);
printf("Address %p\n", argv[0]);
printf("Contents: %s\n", argv[0]);

printf("Pointer Location %p\n", &argv[1]);
printf("Address %p\n", argv[1]);
printf("Contents: %s\n", argv[1]);

if (argc == 3) {
        printf("Pointer Location %p\n", &argv[2]);
        printf("Address %p\n", argv[2]);
        printf("Contents: %s\n", argv[2]);
}
```

5.7 Summary

Arrays and Pointers share a symbiotic relationship. Arrays are simple to learn, but Pointers provide an efficient way to write a program to access data and memory. The discussion on Arrays and Pointers will not end here. The next 4 chapters in the book will use these features to address different computational problems and their solutions.

6 Recursion, Sort, and Search

As more and more data is being generated in the Cloud, the need to efficiently sort and search data to produce useful information becomes a high priority for developers. Searching and sorting data at the device level can focus on effective data being sent to the Cloud in the first place. For example, a device could monitor temperature every 20 minutes. A full day of data can add up quickly when all that is needed is the high, low, and mean values for a day or a week. In this chapter, we will look at algorithm analysis as related to sorting and searching data. Before we go further, we need to talk about a method for solving problems that is used in both sorting and searching.

6.1 Recursion

Recursion is a method to solve a problem using smaller instances of the same problem, or more simply put, to have a function call itself.

The various ways to implement recursion are complex enough that one could dedicate a whole book to the topic. This will be an introductory examination of some of the basic uses of recursion, and there is a reference at the end of the book for those who wish to dive deeper into this topic. Basically, there are two types of recursion: direct recursion and indirection recursion. Direct recursion, as it indicates, is when a function calls itself. Indirect recursion is when a function calls other functions that then call the original function.

The technique of recursion is used in sort and search algorithms. The best way to see recursion in action is to look at some simple exercises.

6.1.1   Computer Activity 6.1 - Factorial Recursion Style

Back in Chapter 3, we covered the Factorial. A factorial is a positive integer n, denoted by "n!", that is a product of all positive integers less than or equal to n. For example, 4! would be 24 = 4 * 3 * 2 * 1. The factorial function is defined as:

$$n! = \prod_{k=1}^{n} k$$

Another way to present the equation n! = n * (n-1)!, thus 5! = 5 * 4!, 4! = 4 * 3!, etc. Of course, 0! = 1. In Chapter 3, we use an If-condition and a for-loop to find a factorial. Recursion is also an iterative loop that has a function that calls itself. We will discuss loops versus recursion a little later. Let's see what the solution looks like using recursion. Create a new C Project in Visual Studio and name the project CH6-FactorialRecursion. Enter the following code:

```c
#include <stdio.h>
#include <string.h>
#include <stdlib.h>

int fact(int n) {

        if (n == 1 || n == 0) {
                return 1;
        }
        else {
                return n * fact(n - 1);
        }
}


void main() {

        char userNumber[10];

        printf("Calculate a factorial\n");
        printf("Enter an integer value for n\n");
        scanf_s("%9s", userNumber, 10);

        int n = atoi(userNumber);

        printf("The factorial for %d is %d\n", n, fact(n));

        for (;;);
}
```

The main() function allows the user to enter a number for n and then print the result of the recursive call to fact(). Within the fact() function, we first check for the base case, which is 1! or 0!. If n is > 1, fact() function is called again. fact() will continuously be called until n == 1, which is the base case. Run the application and enter 5 in the console, and you should see that 120 is the answer.

```
Calculate a factorial
Enter an integer value for n
5
The factorial for 5 is 120
```

6.1.2 Computer Activity 6.2 - Iteration versus Recursion
Recursion is an elegant and clean way to solve certain types of problems that require iterative operations, but there are times when recursion can be inefficient. Each time a method calls a method, it is placed on the stack when the call is made. In a deeply recursive call, the stack usage can take up valuable computer memory resources. For this next example, we will use the Fibonacci sequence:

$$f_1 = 1, f_2 = 1, f_n = f_{n-1} + f_{n-2}$$

The sequence gets the sum of the first prior two numbers in the sequence: 1, 1, 2, 3, 5, 8... Create a new C Project in Visual Studio and name the project CH6-FibonacciRecursion. Enter the following code:

```
#include <stdio.h>

long fibonacci(int n) {

    if (n <= 2) {
        return 1;
    }
    else {
        return (fibonacci(n - 1) + fibonacci(n - 2));
    }
}


void main() {

    for (int i = 1; i <= 45; i++) {
        printf("Fibonacci(%d) result is %ld \n", i, fibonacci(i));
    }

    for (;;);

}
```

The for-loop enters the values 1 through 45 for n. The printf() function makes the call to the recursive method and will output the result for each value of n. Run the application, and you should get the following results:

```
Fibonacci(1) result is 1
Fibonacci(2) result is 1
Fibonacci(3) result is 2
Fibonacci(4) result is 3
Fibonacci(5) result is 5
Fibonacci(6) result is 8
:
:
:
Fibonacci(43) result is 433494437
Fibonacci(44) result is 701408733
Fibonacci(45) result is 1134903170
```

The output for each value of n starts out really fast; but as the value of n gets larger, the output slows down. Create a new C Project in Visual Studio and name the project CH6-FibonacciIteration. Enter the following code:

```c
#include <stdio.h>

long fibonacci(int n) {

        if (n <= 2) {
                return 1;
        }
        else {

                long fn = 0;
                long fn1 = 1;
                long fn2 = 1;

                for (int x = 3; x <= n; x++) {
                        fn = fn1 + fn2;
                        fn2 = fn1;
                        fn1 = fn;
                }
                return fn;
        }
}
```

```
void main() {

    for (int i = 1; i <= 45; i++) {
        printf("Fibonacci(%d) result is %ld \n", i, fibonacci(i));
    }

    for (;;);
}
```

The same base case of n <= 2 is addressed. The for-loop starts the iteration at 3 since n > 2. The equation is used to add the two previous values in the sequence, and two variables hold the previous two values to be used in the next iteration. Run the application and you will notice the output is much faster. Why is the recursive solution slower? We need to map the recursive calls to get a better understanding of what the program execution looks like. Let's look at the case when n = 5 in the figure below There are two recursive calls, we will map the recursive calls to the base case of n <= 2 for clarity.

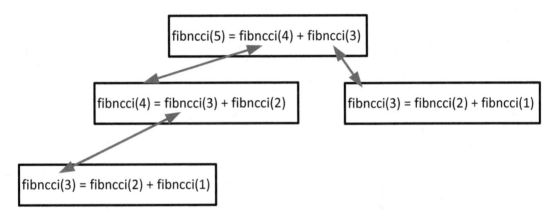

What you see is that multiple calls are being made to the recursive method with the same value, i.e., fibncci(3) is called twice and fibncci(2) is called three times. The greater the values of n, the more repetitions like these occur. From a code execution perspective, the problem with the recursive Fibonacci sequence is the two recursive calls fanning out into more recursive calls. The iterative version of the Fibonacci sequence avoids the repetitious calculations of the recursive version. In this case, the iterative solution is the better solution from a code execution perspective.

6.1.3   Computer Activity 6.3 - Mutual Recursion
The last two examples show a single integer sequence or data set. What if two data sets are dependent on each other? Mutual Recursion is a form of recursion where two functions call each other. Each call is reduced to a simpler set until a base case and the

solution is reached. The most popular mutual recursion example on the Internet is odd and even numbers.

        Odd = 1, 3, 5, 7, 9....
        Even = 2, 4, 6, 8, 10...

Create a new C Project in Visual Studio and name the project CH6-OddEven. Enter the following code:

```
#include <stdio.h>
#include <stdlib.h>

char* check_odd(int n);
char* check_even(int n);

void main() {
        char userNumber[10];

        printf("Enter an integer value\n");
        scanf_s("%9s", userNumber, 10);
        int n = atoi(userNumber);

        printf("The number %d is %s\n", n, check_even(n));

        for (;;);
}


char* check_odd(int n) {

        if (n == 0) {
                return "Odd";
        }
        else {
                return check_even(n - 1);
        }
}

char* check_even(int n) {

        if (n == 0) {
                return "Even";
        }
        else {
                return check_odd(n - 1);
        }
```

}

The main() function accepts user input for a number, and then check_even() is called. The base case of n == 0 is checked. If the base case is not reached, the check_odd() function is called with n decremented by 1. In the check_odd() function, the base case is again checked. If the base case is not reached, the check_even() function is called with n decremented by 1. The calls proceed back and forth between the two methods until n == 0.

6.1.4   Summary of Recursion

At the point of each recursive call, the current state of execution is saved on the stack, and then a jump is made back to the routine with the newly calculated arguments. This takes time in the CPU cycles required to save the state and execute the call, and it takes memory space to save the state and create a new set of local variables. Deeply recursive calls take time and resources, and you run the risk of stack overflow.

6.2   *Algorithm Analysis Introduction – Big-O Notation*

The goal of any algorithm is to complete that task in the shortest time possible in whatever application it is used. The performance of any algorithm is not just determined by how efficiently it is coded, but also by the data sets on which it is being used. The more efficiently the algorithm can handle large amounts of data the better. Algorithm analysis studies the amount of resources in time and storage that an algorithm requires when running. For our introduction, we will focus on time resources and limit our examples to a single data set. Computer Activity 9.2 compared the performance of the different Fibonacci sequence algorithms and showed that the iteration algorithm was much faster than the recursion algorithm. The two Fibonacci algorithms show that there are two ways to solve the problem, but one is more efficient for a small dataset. What happens when the data set gets larger? How an algorithm's performance changes as the data set gets larger is referred to as how well the algorithm scales; that is to say, how does the algorithm's runtime change with the size of the data set it is processing? Comparing algorithms specifically by absolute time measurement is not practical, since different computers and programming languages can change the raw speed of an algorithm. What can be determined is the relative growth rate. This type of analysis can show asymptotic upper and lower bounds, as well as, worst-case and best-case scaling for a given algorithm as the dataset gets larger. The asymptotic bounding methodologies are known as Big Omega-Ω, Big Theta –Θ, and Big – O.

We will use a few diagrams to describe Big-Ω (Big-Omega), Big–Θ (Big-Theta), and Big–O. The use of these methodologies is to provide a lower bound on the execution time of an algorithm, an upper bound on the execution time of an algorithm, or tight bounding of the execution time of an algorithm as the data set size varies in size. Big-O notation gives an asymptotic upper bound on execution time, Big-Ω notation provides an asymptotic lower bound on execution time, and Big–Θ provides asymptotic tight bounding (a hard upper and lower bound) on execution time. Let's examine each of these notation methodologies to see how they work.

The first diagram below shows a complex function f(x), which could be an actual measurement of an algorithm's execution time versus the size of the data set being processed. Next to that function is a simpler function g(x) that has a similar growth rate, especially in the region where the data set being processed is large.

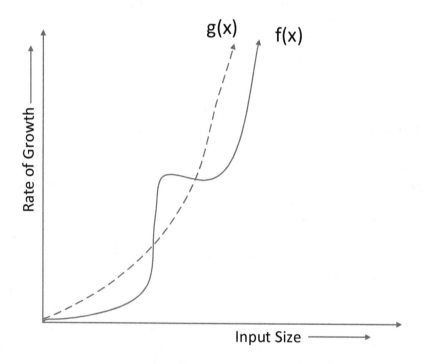

Since Big-Ω notation is a lower bound, the figure below shows that the appropriate constant multiplier, c_2, $c_2*g(x)$ can be chosen such that f(x) will never cross below $c_2*g(x)$ after x_0. This would provide the Big-Ω notation lower bound for f(x).

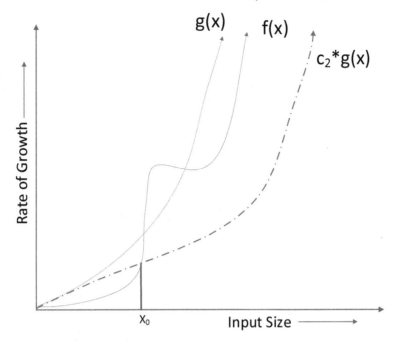

The formula Big-Ω definition would then be:

$$f(x) \in \Omega(g(x)) \text{ if } c_2, x_0 > 0 \text{ such that } f(x) \geq c_2 * g(x) \text{ for all } x \geq x_0$$

Note: the $\in$ symbol above means "is in", such as $f(x)$ is in Big-Ω of $g(x)$ if the following conditions are met.

Big-O notation is the upper bound. **Error! Reference source not found.** figure below shows that for a different constant multiplier, c_1, $c_1 * g(x)$ can be chosen such that $f(x)$ will never cross above $c_1 * g(x)$ after x_0. This would provide the Big-O notation upper bound for $f(x)$.

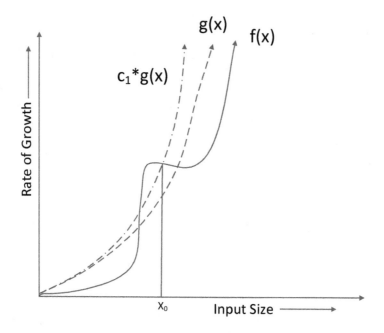

The formula Big-O definition would then be:

$$f(x) \in O(g(x)) \text{ if } c_1, x_0 > 0 \text{ such that } f(x) \leq c_1 * g(x) \text{ for all } x \geq x_0$$

Big-Θ notation is the tight bounding by Big-Ω and Big-O as shown in the figure below. Big-Θ is defined as two constants that $f(x)$ never goes above $c_1 * g(x)$ or below $c_2 * g(x)$ for some point x_0 and after.

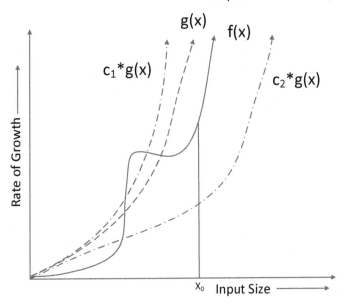

The formula Big - Θ definition becomes:

$$f(x) \in \Theta\ (g(x))\ \text{if}\ c_1,\ c_2,\ x_0 > 0\ \text{such that}\ c_1 {*} g(x) \geq f(x) \geq c_1 {*} g(x)\ \text{for all}\ x \geq x_0$$

Out of the three, Big-O is the one that gets the most attention, since it is the upper bound and has the highest growth rate. You may have read on the Internet or heard from others that understanding Big-O notation involves complicated math, empirical tests, and voodoo. We can set your mind to rest that there is no voodoo required, but there are some high-level math concepts. Understanding the properties of linear, power series, exponential, logarithmic, and other types of functions is required, which are best left to a data structures and algorithms course. For this discussion, sort and search will be examined to demonstrate the basic analysis approach for Big-O notation. In general, a strong foundation in mathematics will help to understand Big-O notation.

6.2.1   A Simple Example
Throughout these examples, 'n' will represent an integer which is the size of the dataset that the algorithm is working on, thus "n" will be used for the Big-O notation. Let's say we have the following function:

$$f(n) = 79n^3 + 2n - 1$$

For larger values n, or as n-> ∞, the contribution that 2n-1 makes to the value of f(n) is relatively small and can be dropped from the Big-O analysis. The $79n^3$ term has the greatest impact on the value of f(n), and hence the growth rate. The constant, 79, can be dropped since it has no bearing on growth rate, thus the Big-O notation for the example function is $O(n^3)$.

Note: In mathematics, "x" is typically designated for real numbers while "n" is for integers.

6.2.2   Code Example of O(1)

The first code example is of a constant output. The code below has a function called AddConstant. No matter how many times the function is called, the time to calculate the result is the same, thus the Big-O notation for a constant is O(1). Notice the array is passed to the function as a pointer (pass by reference).

```c
#include <stdio.h>

void AddConstant(int* anarray, int x) {

        *anarray += x;
}


void main() {

        int myarray[] = { 6,7,8,2,3 };

        AddConstant(myarray, 3);

        for (int i = 0; i < 5; i++) {

                printf("%d  ", myarray[i]);
        }

        printf("\n");
        for (;;);
}
```

6.2.3 Code Example of O(n)

In the code example below, a for-loop is used to add a constant value to all elements in the array. The array is again passed to the function as a pointer. The pointer is then used to perform the arithmetic on each element in the array.

```c
#include <stdio.h>

void AddConstant(int* anarray, int x) {

        for (int i = 0; i < 5; i++) {
                *anarray += x;
                anarray++;
        }
}


void main() {

        int myarray[] = { 6,7,8,2,3 };

        AddConstant(myarray, 3);

        for (int i = 0; i < 5; i++) {

                printf("%d  ", myarray[i]);
        }

        printf("\n");
        for (;;);
}
```

If we pass larger and larger arrays to the AddConstant() function, the time to process the array goes up linearly. How do you figure out the Big-O notation? The for-loop in AddConstant will have "n" elements from the array, and we will assign this line "n". The addition of the value x is constant, and we will assign it 'c'. The total time is T(n) = c*n, thus the Big-O notation is O(n), since the constant does not contribute to the relative rate.

6.2.4 Code Example of O(n²)

The final code example below has the method, ComplexChange(), which features nested for-loops. The application performs an arithmetic computation on each element in the array 5 times. The outer loop controls how many times the arithmetic is performed. The inner loop transverses each element in the array.

```c
#include <stdio.h>
```

123

```
void ComplexChange(int* anarray, int x) {

        int* temp = anarray; //Pointer to a pointer

        for (int i = 0; i < 5; i++) {

                anarray = temp; //Restore the anarray pointer to the first
element in the array

                        for (int j = 0; j < 5; j++) {
                                *anarray = (* anarray * x) - i;
                                anarray++;
                        }
                }
        }
}


void main() {

        int myarray[] = { 6,7,8,2,3 };

        ComplexChange(myarray, 2);

        for (int i = 0; i < 5; i++) {

                printf("%d  ", myarray[i]);
        }

        printf("\n");
        for (;;);
}
```

Like the previous O(n) example, we can assign 'n' for the two for-loops and a constant 'c' for the arithmetic in the inner loop. The total time is $T(n) = c*n*n$ or $T(n) = c*n^2$, thus dropping the constant the Big-O notation is $O(n^2)$. If there were 3 nested for-loops, the Big-O notation would be $O(n^3)$, if there were 4 nested for-loops, the Big-O would be $O(n^4)$, and so on.

6.2.5   The Beginning of the Big Picture
Taking the three code examples, we draw the three growth rates on a single graph.

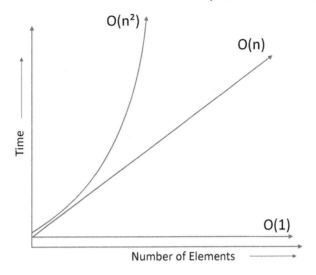

Obviously, $O(n^2)$ has the worst efficiency as the dataset grows. $O(n)$ increases linearly as the dataset grows, and $O(1)$ has a constant time result irrespective of the data set size. Now that we have a basic notation to describe algorithm efficiency, we can dive into sort and search.

6.3   Sort and Search Algorithms

There are many sorting and search algorithm solutions that have been developed over the years. We are not going to cover all of them. Each algorithm has different performance characteristics that are quantified using the Big-O notation. The table below lists some of the fundamental Sort and Search algorithms taught in computer science. The simple Sort and Search algorithms take longer as the data set gets bigger. The more complex Sort and Search algorithms that implement a divide and conquer with recursion have faster performance but take up more system resources.

Algorithm	Big-O
Selection Sort	$O(n^2)$
Insertion Sort	$O(n^2)$
Merge Sort	$O(nlogn)$
Quicksort	$O(nlogn)$
Linear Search	$O(n)$
Binary Search	$O(\log n)$

There is a website (http://bigocheatsheet.com/) that provides the Big-O notation for many popular sort and search algorithms. The following graph shows each Big-O notation.

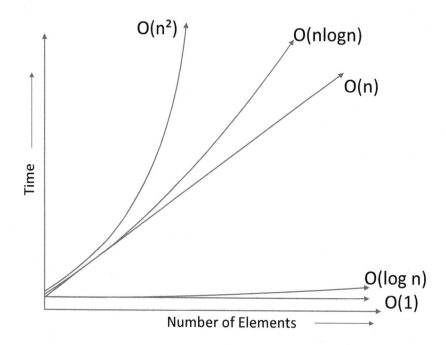

Most computer science textbooks walk through all the algorithms and implementations. The detailed discussion is beyond the scope of this book. If you are interested in learning the implementation details, please see our other book *Java and Eclipse for Computer Science*. From the table, Quicksort (O(nlogn)) and Binary Search (O (log n)) are the best-in-class sort and search algorithms. The <stdlib.h> library contains the Quicksort, qsort(), and Binary Search, bsearch(), functions, which is a reason why we are not going into all the details. There is a saying that 'just because you have a hammer doesn't make every problem a nail'. If you are in a memory-constrained device like an MCU or FPGA, the simpler sort and search algorithms will be better for smaller datasets.

6.3.1   Computer Activity 6.4 – Insertion Sort vs Qsort()
The program will compare the Insertion sort algorithm versus the built-in Qsort function. The CPU timer ticks will be used to calculate the total time to complete the sort. Create a new C Project in Visual Studio and name the project CH6-Sort. Enter the following code:

```
#include <stdio.h>
#include <stdlib.h>
#include <time.h>
```

```
void insertionSort(int datatoSort[], int n) {

    for (int x = 1; x < n; x++) {
        int k = x - 1;
        int temp = datatoSort[x];

        while ((k >= 0) && (datatoSort[k] > temp)) {
            datatoSort[k + 1] = datatoSort[k];
            k--;
        }
        datatoSort[k + 1] = temp;
    }
}

int sortcmp(const void* a, const void* b) {

    return (*(int*)a - *(int*)b);
}

void arrayPrint(int* arrayptr, int size) {

    for (int x = 0; x < size; x++) {
        printf("%d, ", *arrayptr++);
    }
}


void main() {

    clock_t starttime, endtime;
    double totaltime;

    int n = 100;
    int raDS1[100];
    int raDS2[100];
    srand((unsigned int)time(NULL));
    printf("Insertion versus Qsort\n");
    printf("here is the array\n");
    for (int x = 0; x < n; x++) {
        raDS1[x] = raDS2[x] = (int)rand() % 100; //Duplcate the
same values
    }
    arrayPrint(raDS1, n);
    printf("\n");

    //Insertion Sort
    printf("\n");
    printf("Insertion sort\n");
```

```
        starttime = clock();
        insertionSort(raDS1, n);
        endtime = clock();

        totaltime = (double)(endtime - starttime) / CLOCKS_PER_SEC;

        printf("Here is the sorted array\n");
        arrayPrint(raDS1, n);
        printf("\n");
        printf("Insertion sort total time %lf\n", totaltime);

        printf("\n");
        //Qsort
        printf("Qsort\n");
        starttime = clock();
        qsort(raDS2, n, sizeof(int), sortcmp);
        endtime = clock();

        totaltime = (double)(endtime - starttime) / CLOCKS_PER_SEC;

        printf("Here is the sorted array\n");
        arrayPrint(raDS2, n);
        printf("\n");
        printf("Qsort total time %lf\n", totaltime);

        for (;;);
}
```

There are three functions declared and defined before main(). The insertionSort() function is an implementation of the insertion sort algorithm. The sortcmp() function is used by the qsort() function to compare two values and return a result. The result is either a > b, a = b, or a < b. The arrayPrint() function prints out all elements of the array. In the main(), two duplicate arrays are filled with random integers 0-99. The Insertion Sort is called first. The start and end times are collected before and after calling the insertionSort() function. The sorted array is then printed to the screen along with the total time it took to run the function. The qsort() function is called using the duplicate array. Again, a start and end time is collected. The sorted array is then printed to the screen along with the total time it took to run the function.

1. Run the program. The program prints out the unsorted array and you can see that both sorting algorithms output the same result and similar time.
2. Comment out all three arrayPrint() function calls. We need to do this since an increase in the number of elements will fill up the command window.

3. Change n to equal 2000 and change the array to take in 2000 elements raDS1[2000] and raDS2[2000]
4. Run the program. The times are still the same.
5. Change n =100000 and raDS1[10000] and raDS2[10000]
6. Run the program again. The Insertion Sort is a little slower.
7. Change n =500000 and raDS1[50000] and raDS2[50000]
8. Run the program. The insertion sort gets a little slower while the qsort() function stays the same.

```
Insertion versus Qsort
here is the array


Insertion sort
Here is the sorted array

Insertion sort total time 0.858000

Qsort
Here is the sorted array

Qsort total time 0.003000
```

If you are dealing with a small amount of data, both algorithms perform at the same speed; but as the data set gets bigger, the qsort() function is much quicker. Again, using the right algorithm for the right computer system is an important consideration.

6.4   Computer Activity 6.5 – Linear versus Binary Search

There are two elementary search functions that are covered in computer science courses: Linear and Binary. The Binary by Big-O notation is faster as the number of elements gets bigger. The <stdlib.h> library has the bsearch() function built in. The program we create will compare the two algorithms, but because we are writing in C on a PC platform, you will not see a big difference in performance. Create a new C Project in Visual Studio and name the project CH6-Search. Enter the following code:

```c
#include <stdio.h>
#include <stdlib.h>
#include <time.h>

int linearSearch(int datatoSearch[], int n, int data) {

        for (int i = 0; i < n; i++) {
```

```c
            if (datatoSearch[i] == data) {
                    return i;
            }
        }
        return -1; //Data not found
}
int compare(const void* a, const void* b) {

        return (*(int*)a - *(int*)b);
}


void arrayPrint(int* arrayptr, int size) {

        for (int x = 0; x < size; x++) {
                printf("%d, ", *arrayptr++);
        }
}


void main() {

        clock_t starttime, endtime;
        double totaltime;

        int n = 100;
        int raDS1[100];
        int raDS2[100];
        int searchValue = 76;
        int founddata;
        srand((unsigned int)time(NULL));
        printf("Linear versus Binary\n");
        printf("here is the array\n");
        for (int x = 0; x < n; x++) {
                raDS1[x] = raDS2[x] = (int)rand() % 100; //Duplcate the
same values
        }
        //arrayPrint(raDS1, n);
        printf("\n");

        //Linear Search
        printf("\n");
        printf("Linear Search\n");
        starttime = clock();
        founddata = linearSearch(raDS1, n, searchValue);
        endtime = clock();
        totaltime = (double)(endtime - starttime) / CLOCKS_PER_SEC;
        if (founddata != -1) {
                printf("76 was found\n");
        }
```

```
        else {
                printf("76 was not found\n");
        }

        printf("Linear search total time %1f\n", totaltime);
        printf("\n");


        //binary search
        printf("Binary\n");
        starttime = clock();
        qsort(raDS2, n, sizeof(int), compare); //Sort the array first
        founddata = bsearch(&searchValue, raDS2, n, sizeof(int), compare);

        endtime = clock();
        totaltime = (double)(endtime - starttime) / CLOCKS_PER_SEC;
        if (founddata != -1) {
                printf("76 was found\n");
        }
        else {
                printf("76 was not found\n");
        }
        printf("Binary search total time %1f\n", totaltime);

        for (;;);
}
```

Like the Sort program, duplicate arrays are filled with random integers. The Linear search walks through each element in the array, and if the desired number is found, returns the index where it was found. The bsearch() function requires the qsort() function to be called first, and the bsearch() can look for the value in the sorted array. A Linear search doesn't need sorting. Run the program and the timing for each will be the same. Increase the number of elements, and the bsearch() will take a tiny bit longer because we included the qsort() function in the timing. Keep in mind that on memory-constrained systems with small data sets, Linear search would be a preferred choice.

6.5 Summary

The chapter covers the concepts of Recursion, Sort, and Search. Recursion is a powerful programming method and will be an asset to your programming skill set. The key point is to break the problem down to the base case, which will provide the kick to derive the result. Recursion is used in specific situations, such as sorting and searching where it becomes a little complex for iterative solutions. Recursion makes heavy use of the stack, so be careful in your use of recursion algorithms and your program memory layout that you do not overflow the stack. The Mergesort algorithm takes advantage of recursion. We didn't go into the mathematics or implementation of all the different sort and search algorithms, but computer science textbooks cover these functions in detail. Choosing the right algorithm for your application and system is important. It is nice that the qsort() and bsearch() are available in the C library, but they might not be the best fit for all scenarios. For systems with limited memory like MCUs, you have to pick the right solution that fits with the available resources. Please keep the other algorithms in mind as you design your system or even consider crafting your own algorithm.

7 Create Complex Data Types

All the projects so far have used the built-in data types and the array data structures. In this chapter, we will look at creating different complex data types by combining built-in data types to address more complex data problems.

7.1 *Typedef*

The typedef keyword is used to rename a data type into something more useful. For example, managed code languages like C# and Java have a "string" data type. In C, you have char[] or *char. With typedef, you can define string as a char pointer.

    typedef char* String

Now "String" can be used as a data type. The #define was discussed in Chapter 4, and it performed a similar function, but the difference is that #define is a preprocessor directive, whereas, typedef is processed by the compiler. Also, typedef is restricted to only renaming data types.

7.1.1 Computer Activity 7.1: String Data Type

Create a new C project in Visual Studio called CH7-StringType. Enter the following code:

```c
#include <stdio.h>

void main() {

    typedef char* String;

    String mystring = "Hello World";

    printf("%s\n", mystring);

    for (;;);
```

```
}
```

Run the program and the message is displayed in the console. If you put the typedef in a header file, you can include the header file in the application and just use the String data type to define strings.

7.2 Struct

A structure is a collection of related variables that define a data type. For example, an address book application that stores name, address, phone, etc. might have a structure defined as follows:

```
struct mailaddress {
        char contactName[20];
        char contactAddress[30];
        char contactCityStZip[50];
        char contactPhone[16];
        char contactEmail[30];
};
```

The individual arrays in the structure are called structure members. The mailaddress structure has members of the same data type, but a structure can have members with different data types or other structures. Now that we have a structure data type defined, we can declare an array using the structure:

```
struct mailaddress mycontacts[1000];
```

Members of a structure can be passed to a function, or the full structure can be an argument to a function. Alternatively, if the structure is only going to be declared one time, the array can be placed at the end of the structure definition:

```
struct mailaddress {
        char contactName[20];
        char contactAddress[30];
        char contactCityStZip[50];
        char contactPhone[16];
        char contactEmail[30];
} mycontacts[1000];
```

7.2.1 Computer Activity 7.2: Simple Structure

The program will demonstrate accessing members in a structure. Create a new C project in Visual Studio called CH7-SimpleStructure. Enter the following code:

```c
#include <stdio.h>

void main() {

    struct Coordinates3d {

        int x_axis;
        int y_axis;
        int z_axis;
    } pointlocation;

    pointlocation.x_axis = 7;
    pointlocation.y_axis = 10;
    pointlocation.z_axis = 2;


    printf("The point is located at (%d,%d,%d)\n",
pointlocation.x_axis, pointlocation.y_axis, pointlocation.z_axis);

    for (;;);
}
```

Run the application and the values assigned to each member in the structure are printed in the string output. The dot "." notation is used in combination with the declared structure's name and the member's name to set and get member data.

7.2.2 Computer Activity 7.3: Pointers to a Structure

Pointers can be used to access structure members. Create a new C project in Visual Studio called CH7-PointerToStructure. Enter the following code:

```c
#include <stdio.h>

void main() {

    struct Coordinates3d {

        int x_axis;
        int y_axis;
        int z_axis;
```

```
        } pointerlocation = {7,10,2};

        struct Coordinates3d* myptr;
        myptr = &pointerlocation;

        printf("The point is located at (%d,%d,%d)\n", myptr->x_axis,
    myptr->y_axis, myptr->z_axis);

        for (;;);
    }
```

Run the application, and you will get the same result. The pointerlocation declaration includes the assignment for all three structure members. A pointer with the structure data type is created. The dash greater-than combination "->" is used to access members of the structures using the structure pointer. Working with structures can get a little confusing between the dot "." and the arrow "->".

7.2.3   Computer Activity 7.4 Passing a Structure to a Function
Let's see what happens when we pass a structure to a function. Create a new C project in Visual Studio called CH7-StructurePassing. Enter the following code:

```
#include <stdio.h>

struct Coordinates3d {

        int x_axis;
        int y_axis;
        int z_axis;
};

void changeCoordinates(struct Coordinates3d pointloc) {

        pointloc.x_axis += 5;
        pointloc.y_axis -= 6;
        pointloc.z_axis *= 4;

        printf("The point is located at (%d,%d,%d)\n", pointloc.x_axis,
    pointloc.y_axis, pointloc.z_axis);

}

void main() {
```

```
    struct Coordinates3d pointerlocation = { 7,10,2 };

    changeCoordinates(pointerlocation);

    printf("The point is located at (%d,%d,%d)\n",
pointerlocation.x_axis, pointerlocation.y_axis, pointerlocation.z_axis);

    for (;;);
}
```

The structure is defined outside the main() as it will be used as input arguments for a function. Run the application and you will get the following output:

```
The point is located at (12,4,8)
The point is located at (7,10,2)
```

Passing the whole structure is simpler than passing each member. The structure is a data type so the members were passed as arguments. The changeCoordinates() function changed the values locally and printed the result. When the application returned to main, the local values assigned in main() have not changed.

7.2.4   Computer Activity 7.5: ArrayofStructures
The program will demonstrate an array of structures. Building on the last computer activity a pointer to the structure will be passed to a function to change a specific member value. Create a new C project in Visual Studio called CH7-StructArray. Enter the following code:

```
#include <stdio.h>

struct inventory {
        char* product;
        int qty;
};

void qtychange(struct inventory *local) {

        local->qty = 10;
}
```

```
void main() {

        struct inventory mystock[3];
        struct inventory* mystockptr;
        mystockptr = mystock;

        mystock[0].product = "Soap";
        mystock[0].qty = 15;

        mystock[1].product = "toothpaste";
        mystock[1].qty = 20;

        mystock[2].product = "shampoo";
        mystock[2].qty = 8;

        mystockptr++;
        qtychange(mystockptr);

        for (int x = 0; x < 3; x++) {
                printf("There is %d quantity of %s\n", mystock[x].qty,
mystock[x].product);
        }

        for (;;);
}
```

A structure called inventory contains the product name and the quantity in stock. An array is declared using the structure data type, and then a pointer is created to point to the structure. The members of each structure in the array of structures are then set to their respective values. The pointer to the array of structures points to the first product in the array. The pointer is incremented to point to the second product structure and passed to the qtychange() function, where the qty value of the second structure in the array of structures is changed. Run the code and you will see that the toothpaste quantity was changed from the initial value of 20 to 10. Where the previous computer activity passed the structure values as arguments, the pointer passes the address so the function can modify the data addressed by the pointer. Passing an argument to a function by value creates a copy of that value within the function. Anything done to or with that value is strictly local to the function the original data used in the call is unchanged. Passing an argument by reference, passing a pointer, passed the address of that argument to the function. Since the function then has the address of that argument it is free to not only get the value of that argument and use it, but it can also modify the original argument

138

that was used in the call. The difference between passing by value and passing by reference is subtle but important.

7.3 Bit Fields

A special structure called a Bit Field defines a set of continuous bits. For example, an 8-bit register for a UART in an MCU will contain bits for baud rate, data bits, parity, and flow control. A bit field structure to set and get the individual bits would look like the following:

```c
struct uartconfig {

        unsigned buad : 3;
        unsigned databits : 1;
        unsigned parity : 2;
        unsigned flow : 2;
};
```

The MCU datasheet would define the actual settings for each of the bit combinations. Here is a full code example:

```c
#include <stdio.h>

struct uartconfig {

        unsigned buad : 3;
        unsigned databits : 1;
        unsigned parity : 2;
        unsigned flow : 2;
};

void main() {

        struct uartconfig uart1 = {7,1,0,0};

        char* baud ="";
        char* databits = "";
        char* parity ="";
        char* flow="";

        switch (uart1.buad) {
                case 0:
                        baud = "4800";
                        break;
                case 1:
```

```
                        baud = "9600";
                        break;
                case 2:
                        baud = "19200";
                        break;
                case 3:
                        baud = "28800";
                        break;
                case 4:
                        baud = "38400";
                        break;
                case 5:
                        baud = "57600";
                        break;
                case 6:
                        baud = "76800";
                        break;
                case 7:
                        baud = "115200";
                        break;
        }

        switch (uart1.databits) {
                case 0:
                        databits = "7";
                        break;
                case 1:
                        databits = "8";
                        break;
        }

        switch (uart1.parity) {
                case 0:
                        parity = "None";
                        break;
                case 1:
                        parity = "Odd";
                        break;
                case 2:
                        parity = "Even";
                        break;
                case 3:
                        parity = "None";
                        break;
        }

        switch (uart1.flow) {
```

```
            case 0:
                    flow = "None";
                    break;
            case 1:
                    flow = "RTS/CTS";
                    break;
            case 2:
                    flow = "XON/XOFF";
                    break;
            case 3:
                    flow = "Hardware";
                    break;
    }


    printf("Uart1 settings are:\n");
    printf("Baud Rate: %s\n", baud);
    printf("Data Bits: %s\n", databits);
    printf("Parity:    %s\n", parity);
    printf("Flow control: %s\n", flow);

    for (;;);
}
```

The output is:

```
Uart1 settings are:
Baud Rate: 115200
Data Bits: 8
Parity:    None
Flow control: None
```

Bit fields are tricky and aren't the most efficient way of working with bit-wise structures. Bits of a bit field aren't always stored as single hardware bits. A single bit could be stored in one full integer. The physical storage and ordering of bits are hardware and compiler-dependent. Bit fields are not arrays and don't have addresses; therefore, the & operator cannot be applied to them. Because bits don't have addresses, you cannot directly overlay the bit structure on a hardware device with memory-mapped registers.

7.4 *Enumeration*

Enumeration is a user-defined list of constant integer values. By default, each constant in the numeration is assigned an integer value starting with 0, but the value can be overridden. In the example below, an enumeration is defined for week. Since enumerations are automatically set to start with 0, the Sunday = 1 overrides the default, and the constant is set to 1. All other constants that follow that are not overridden are incremented by 1. Monday becomes 2, Tuesday is 3, etc. Within the main() function, "day" is declared as an enum week with the value set to Wednesday.

```c
#include <stdio.h>

enum week {Sunday=1, Monday, Tuesday, Wednesday, Thursday, Friday,
Saturday};

void main() {

        enum week day = Wednesday;

        printf("The week day number is %d", day);

        for (;;);

}
```

When the program runs the output is:

```
The weekday number is 4
```

7.5 *Unions*

The last data type is a union. A union is a variable that can hold different variables of different types and storage sizes overlaid on each other in memory. The compiler manages the storage of the union. The total storage size of the union will be determined by the largest variable in the union. The compiler will also manage the alignment of all the variables in the union. Any of the variable types in a union may be used to access the union with the restriction that the type read must be the same as the last type written. Be careful in the usage of the union. It is up to the programmer to use the types in the proper order. Unions are frequently misused for data type conversions, writing one union variable type and then turning around and reading another union variable type. This is
142

technically undefined action, but most compilers, these days, do the right thing. It still is not recommended. The primary use of a union is to save on required memory space in memory-restricted systems. Programmatically, the members of a union are accessed the same way that structure members are accessed by using dot "." notation for a variable or a "->" for a pointer.

7.5.1 Computer Activity 7.6: Unions

Create a new C project in Visual Studio called CH7-StructurePassing. Enter the following code:

```c
#include <stdio.h>

union test {
        int x;
        char* string;
};

void main() {

        union test mytest;

        mytest.string = "Hello World";
        printf("%s\n", mytest.string);

        mytest.x = 7;
        printf("x = %d\n", mytest.x);

        //printf("%s\n", mytest.string);

        for (;;);
}
```

Run the program and you will get the following:

    Hello World
    x = 7

Uncomment the last printf() statement, and run the program. The program crashes. The union was first accessed, written, and then read, using the string pointer. Then it was accessed and written as an integer. This overwrote part of the string pointer. When the union was read as an integer, only the part of the memory storage that was used to write the integer was read as an integer, as it should have been. But when the union was read as a pointer to string right afterwards, the overwritten pointer was no longer a valid pointer into memory, and the program crashed trying to access memory that either didn't exist or the operating system did not allow it to access. The string value, pointer to string, has been overwritten as an integer, so the attempt to read it as a string pointer and print what it is pointing to crashes the system. Keep in mind the compiler doesn't check for the validity of the usage of the union members. With today's modern hardware and readily available memory, there is really no good reason to use a union.

7.6   *Summary*

We can see that it is possible to build on the fundamental data types to create new data types for our programs. Our programs become more useful when the data storage and manipulation of our program conceptually match the related data that we are working with. The ability to combine data types into structures can be helpful when creating a record, which is the foundation of a database. In this chapter, several examples to access and change data in structure members were given. Being able to manage all of this data in the program is great, but once the program is terminated, the data will be lost. The next chapter looks at how we can save this data to files.

8 File IO

C has been ported to many different processors and operating system architectures. A main task for those doing the porting is the File I/O access for the <stdio.h> library. The C template used for this book contains <stdio.h> by default, since it provides the basic printf(), getchar(), putchar(), and scanf() functions that we have used so far to access standard input and output (keyboard and monitor). Since C is a standard, the underpinnings of the function to access the hardware are defined by those doing the port. You don't have to worry about the hardware access details. The library also contains many more functions to access the files. In this chapter, we will explore the file functions.

8.1 *Computer Activity 8.1: Write Strings and Read Strings*

Let's dive right in and create a project that creates a new file, writes a string to the file, closes the file, opens the file, and reads the string from the file. Create a directory called c:\testfiles. Create a new C project in Visual Studio called CH8-FileIO1. Enter the following code:

```c
#include <stdio.h>

void main() {

    FILE* fpr, *fpw;
    errno_t err;

    char *outstring = "This is the test message in the file\n";
    char instring[50];

    err = fopen_s(&fpr,"c:\\testfiles\\mynote.txt", "w");
    if (fpr == NULL) {
        printf("Failed to create file\n");
    }
    else {
        fputs(outstring, fpr);
        fclose(fpr);
```

```
        }

        err = fopen_s(&fpw,"c:\\testfiles\\mynote.txt", "r");
        if (fpw == NULL) {
                printf("Failed to open file to read\n");
        }
        else {

                if(fgets(instring, 50, fpw) == 0)
                {
                        printf("Error getting string");
                }
                else {
                        printf("%s", instring);
                }
                fclose(fpw);
        }

        for (;;);
}
```

Run the program and you will get the contents of the file printed to the screen. The first thing you notice in the code is that there is much more involvement in addressing error handling. The very first line in main() declares two FILE pointers; one for writing and the other for reading. FILE is a special structure (object) that identifies the file stream and a pointer to the buffer. The err variable gets the error returns from some functions and then two arrays are declared. One that has the string to be written to the file and the other that will read in the strings from the file. The fopen_s() addresses the buffer overrun security issue that fopen() has. The fopen_s() sets the fpr pointer to the file stream. The mode the file is opened is important. Since this is a new file, the mode is set to "w", thus the file will be created. The table below lists several modes a file can be opened in.

Mode	Description
"r"	Opens a text file to read. If the file can't be found, the call fails.
"w"	Opens a text file to write. If the file doesn't exist, it will create the file. If the file does exist, the pointer will be set to the start of the file, thus wiping out the data already in the file.
"a"	Opens a text file to write in append mode. If the file doesn't exist, it will create the file. If the file does exist, the pointer will be set to the end of the file, thus you will be adding on to the file.

"r+"	Opens a text file to read and write. If the file can't be found, the call fails.
"w+"	Opens a text file to read and write. If the file doesn't exist, it will create the file. If the file does exist, all the contents are lost.
"a+"	Opens a text file to read and write. If the file doesn't exist, it will create the file. Reading will start from the front, and writing will be appended to the end.
"rb"	Opens a binary file to read. If the file can't be found, the call fails
"wb"	Opens a binary file to write. If the file doesn't exist, it will create the file. If the file does exist, the pointer will be set to the start of the file, thus wiping out the data already in the file.
"ab"	Opens a binary file to write in append mode. If the file doesn't exist, it will create the file. If the file does exist, the pointer will be set to the end of the file, thus you will be adding on to the file.

If the file has been successfully opened, the string is written to the file, and the file is closed. The fclose() function has to be put into an if-then condition since it is possible fopen_s() can fail. Visual Studio will warn you if it detects an unreachable code path. The program opens the file again for reading, and upon a successful open, reads the file and prints the contents to the screen. The fputs() and fgets() functions work with strings. The fputc() and fgetc() functions work with characters. The fgets() reads 50 characters are one shot, but if the file has more than 50 characters, you could read another 50 until the EOF file is reached. A different implementation to read in the file contents is to use fgetc() in a loop until the end of file is reached:

```c
#include <stdio.h>

void main() {

        FILE* fpr, *fpw;
        errno_t err;
        char c;

        char *outstring = "This is the test message in the file\n";
        char instring[50];

        err = fopen_s(&fpr,"c:\\testfiles\\mynote.txt", "w");
        if (fpr == NULL) {
                printf("Failed to create file\n");
        }
        else {
                fputs(outstring, fpr);
                fclose(fpr);
```

```
        }


            err = fopen_s(&fpw,"c:\\testfiles\\mynote.txt", "r");
            if (fpw == NULL) {
                    printf("Failed to open file to read\n");
            }
            else {

                    while (1) {

                            c = fgetc(fpw);
                            if (c == EOF) {
                                    break;
                            }
                            else
                            {
                                    printf("%c", c);
                            }
                    }

                    fclose(fpw);
            }

        for (;;);
}
```

8.2   Computer Activity 8.2: Saving a Data Structure

Strings, characters, numeric data types, and arrays are simple to save to a text file; but per the last chapter, most data is a complex mix. The last chapter covered structures, which combined related data types. The program for this computer activity will store several records of a data structure to a file. If you have not done so already, create a directory called c:\testfiles. Create a new C project in Visual Studio called CH8-FileIO2. Enter the following code:

```
#include <stdio.h>
#include <string.h>

struct inventory{

        char prodName[20];
        int qty;
};
```

```
void main() {

        FILE* fp;
        struct inventory invProduce[4], invProduceRead;
        errno_t err;
        fpos_t pos=0;
        int count = 0;

        strcpy_s(invProduce[0].prodName, sizeof(invProduce[0].prodName),
"Apples");
        invProduce[0].qty = 40;

        strcpy_s(invProduce[1].prodName, sizeof(invProduce[1].prodName),
"Oranges");
        invProduce[1].qty = 30;

        strcpy_s(invProduce[2].prodName, sizeof(invProduce[2].prodName),
"Bananas");
        invProduce[2].qty = 55;

        strcpy_s(invProduce[3].prodName, sizeof(invProduce[3].prodName),
"Lettuce");
        invProduce[3].qty = 23;

        err = fopen_s(&fp, "c:\\testfiles\\inventory.dat", "wb");
        if (fp == NULL) {
                printf("Failed to create file\n");
        }
        else {

                for (int x=0; x < 4; x++) {
                        fwrite(&invProduce[x], sizeof(invProduce), 1, fp);


                }
                fclose(fp);
        }


        err = fopen_s(&fp, "c:\\testfiles\\inventory.dat", "rb");
        if (fp == NULL) {
                printf("Failed to open file\n");
        }
        else {

                while (fread(&invProduceRead, sizeof(invProduce), 1, fp) ==
1) {
```

```
                    printf("Produce Name: %s\n",
 invProduceRead.prodName);
                    printf("Quantity: %d\n\n", invProduceRead.qty);
                    count++;
            }

            printf("\nSearch for a specific item\n");
            fseek(fp, 0, SEEK_SET); //point to the beginning of file
            for (int y = 0; y < count; y++) {
                    fseek(fp, sizeof(invProduceRead) * y, SEEK_SET);
                    fgetpos(fp, &pos);
                    printf("%I64u\n", pos);
                    fread(&invProduceRead, sizeof(invProduce), 1, fp);
                    if (strcmp(invProduceRead.prodName, "Bananas")==0) {
                            printf("Produce Name: %s\n",
 invProduceRead.prodName);
                            printf("Quantity: %d\n\n",
 invProduceRead.qty);
                    }
            }

            fclose(fp);
    }

    for(;;);
}
```

Run the program, and you will get the following output:

```
    Produce Name: Apples
    Quantity: 40

    Produce Name: Oranges
    Quantity: 30

    Produce Name: Bananas
    Quantity: 55

    Produce Name: Lettuce
    Quantity: 23


    Search for a specific item
    0
```

```
24
48
Produce Name: Bananas
Quantity: 55

72
```

The inventory structure has been architected to have a fixed 24-byte size for each record: 20 characters and an integer. An array of the inventory structure is filled with the names and quantities of four types of produce. A binary file is opened, and the fwrite() function writes each record to the file moving the stream pointer to the end of the written record on each write. The file is closed and then re-opened. The fread() function reads each record by the structure's size, thus moving the stream pointer 24 bytes after reading. Each record is printed to the screen.

After the last record has been printed, a search for a specific item is performed. The fseek() locates each record by searching from the beginning by an offset of the inventory structure size plus the y index value. The fgetpos() function is used to show the location of the pointer stream after the fseek() function call.

8.3 *Summary*

The <stdio.h> library provides the functions to store and retrieve data. Rather than going through a background on each function in the library, the computer activities demonstrated different file functions and the requirements to address possible errors. Building on the previous chapter, the second computer activity demonstrated how structures can be stored and retrieved from files. The inventory structure is very small, but structures can be much bigger. A programmer needs to be careful with the size of the data being processed, which brings us to the last chapter memory management.

9 Memory Management

We saved the best topic for last. For those who learn C#, Java, or other managed code languages, the handling of memory is performed by the garbage collector. There is no such garbage collection concept within the C/C++ language. You the programmer, in conjunction with library functions, are responsible for allocating and freeing up memory in a C/C++ program. To do this properly, some careful planning is in order. This chapter will focus on the memory management functions found in the C Standard Library and memory management guidelines will be discussed at the end of the chapter.

9.1 *Memory Layout*

The first step to understanding how to manage memory is looking at a program's memory layout. In an operating system such as Windows or Linux, applications are loaded into memory dynamically, thus the actual address location will be different each time a program runs. When there is no operating system such as the case with an FPGA, the application is loaded in the same place each time. Below is a typical layout for a compiled C program:

Memory Layout

The memory layout is broken into two basic sections Static and Dynamic Memory. The static memory section consists of the following:

- Text/Code Segment – The compiler turns the code into machine-level instructions, which are placed in the lowest address area. These instructions never change; thus, they are static.
- Initialized Data Segment – Any variable or array that has been initialized with a value(s) upon creation is stored in this segment. The program has full read-and-write access to the segment to read and to change these variables.

- Uninitialized Data Segment or BSS Segment. The acronym BSS is old: block started by the symbol. For any variable that has not been initialized with a value will be located in this segment. The program has full read-and-write access to this segment as well. At runtime, the value of any uninitialized variable is not known and is implementation and compiler-specific. The initial values of uninitialized variables should never be counted on or used.

The compiler sets up the three static memory segments, and there is not much to be done programmatically. Dynamic memory is where we need to pay careful attention. The dynamic memory section consists of the following:

- Stack – is a location in memory where local variables and data are temporarily stored and removed in a last-in-first-out order (LIFO). The stack is also used for function call management. When a function is invoked, the function is allocated space on the stack or stack frame. The parameters passed into the function, the local variables in the function, and the return address of the function exist in the stack frame, but the return value is not. A function calling a function, calling a function, results in three functions on the stack. When debugging an application, you may have seen the Call Stack window displaying where the program currently is in the debug session. The active function in the debugger will have a line number that it is presently on. The calling function lists the line number that made the call.

Call Stack	
Name	Language
⇨ CH7-StruturePassing.exe!changeCoordinates(Coordinates3d pointloc) Line 12	C
CH7-StruturePassing.exe!main(...) Line 26	C
[External Code]	

As items come off the stack, the memory is automatically freed up. The Stack has some memory management by design. The infamous stack overflows occur when the application fills up the available stack storage and then asks for more. If the base case is missed in a recursion call sequence, it does not complete and turns into an infinite recursion sequence. A programming bug like this will cause a stack overflow. If a program attempts to copy more data into an array than the array will allow, the stack will be corrupted. The recursion computer activities in Section 6.1 Recursion, make heavy use of the stack and come close to the stack overflow condition. Stack-intensive programs need to be well thought out and built with sufficient stack space to handle the worst-case scenario.

- Heap – The stack has limited memory and there is no control for this section. For larger data sets, the Heap is used. The Heap requires that you programmatically allocate and free the memory. When you make allocation and free calls, the memory being allocated is coming from the heap free pool; and when allocated memory is freed, it is released back to the heap free pool. Chapter 7 showed how to create a mail address structure. If the structure was put into main(), Visual Studio will warn you that there is not enough stack space. The recommendation is to allocate memory from the Heap.

```
void main() {

                void main()

                Search Online

        str     C6262: Function uses '146004' bytes of stack. Consider moving some data to heap.

        char contactAddress[30];
        char contactCityStZip[50];
        char contactPhone[16];
        char contactEmail[30];
    } mycontacts[1000];
```

The memory layout diagram shows the typical relationship of the stack to the heap. They grow in opposite directions and their placement in memory is such that when the stack meets the heap there is no longer any free memory available. In memory-constrained systems, you need to pay close attention to memory usage.

9.2   Heap and Memory Functions

The <stdlib.h> library contains three functions to allocate and one function to free memory in the Heap. The only way to access Heap memory is through a pointer.

Function	Description
malloc(size)	Allocates memory by the size in bytes and returns a pointer to the location in Heap memory.
calloc(n, size)	Allocates a number of elements, sets them to zero, and returns a pointer to the location in Heap memory.
realloc(ptr, size)	Changes the size of the previously allocated block of heap memory.
free(ptr)	Deallocates the memory previously allocated, thereby returning it to free Heap memory.

9.2.1 Computer Activity 9.1: Basic Memory Allocation

Create a new C project in Visual Studio called CH9-BasicHeap. Enter the following code:

```c
#include <stdio.h>
#include <stdlib.h>


void main() {

        int* ptr = (int*)malloc(sizeof(int));
        if (ptr != NULL) {
                *ptr = 17;
                printf("The value stored in the Heap is: %d\n", *ptr);
                free(ptr);
        }
        else {
                printf("Failed to allocate memory\n");
        }

        for (;;);
}
```

The program creates a pointer to an integer. Since the allocation of memory can fail, the if-else statement is used to check for a NULL pointer indicating the allocation failure. In this simple program, the allocation should succeed, and a value of 17 is stored in the allocated memory. The value is then printed to the screen. Once we are done with the ptr, a call to the free() function is made to return the memory to the Heap before the program closes.

9.2.2 Computer Activity 9.2: Array of Values

The sizeof() function was used to set the size of the memory allocated. An integer value of 4 could have been used if the hardware that the compiler creating the code for supports a 4-byte integer. Chapter 5 demonstrated the relationship between arrays and pointers, but we can only access the Heap with a pointer. Let's create a program that stores multiple values in the Heap. Create a new C project in Visual Studio called CH9-MultiValues. Enter the following code:

```
#include <stdio.h>
#include <stdlib.h>


void main() {

        int* ptr = (int *)malloc(12);
        int* ptrstart = ptr; //Preserve the start address
        if (ptr != NULL) {
                *ptr = 17;
                ptr++;
                *ptr = 21;
                ptr++;
                *ptr = 4;

                printf("The current value pointed to in the Heap is: %d\n",
    *ptr);
                ptr--;
                ptr--;
                printf("The current value pointed to in the Heap is: %d\n",
    *ptr);
        }
        else {
                printf("Failed to allocate memory\n");
        }

        if (ptr != NULL) {
                ptr = ptrstart; //Must return ptr to the correct memory
    location before freeing memory
                free(ptr);
        }

        for (;;);
}
```

Run the program and the values for 4 and 17 are printed to the console.

The current value pointed to in the Heap is: 4
The current value pointed to in the Heap is: 17

The malloc() function call sets up a pointer to 12 bytes in Heap memory (4 bytes x 3, or (3 * sizeof(int))). The address of the ptr is pointed to by another pointer. Since allocation of memory can fail the if-else statement is used to check that the allocation succeeded. Next, 3 integer values are stored in memory. The pointer is moved and incremented after each integer value is written. The last value is then printed. Then the ptr is moved back, and decremented, until the pointer is, again, pointing to the first integer in the Heap, and the value is read and printed. In order to free the memory, the pointer passed to the free() function must be pointing to the original first address or the Heap will be corrupted. Visual Studio will perform a hard break on the application. To test this add a ptr++ after the last printf() in the if-else and comment out the ptr = ptrstart, run the program. Visual Studio forces a break.

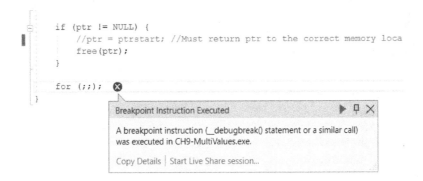

Stop the debug session, uncomment the ptr = ptrstart, and run the program. This time the program runs normally. Technically, Free(ptrstart) would be an alternative to an extra line that resets the initial ptr address. If ptr were to access memory space outside the allocated space, the return values would be garbage. Since we can only access the Heap via pointers, you must keep careful track of the location of the allocated pointers at all times. Preserving the original address on allocation is recommended.

9.2.3 Computer Activity 9.3: Strings

This computer activity will look at strings in Heap memory, as well as, expand the allocated Heap to add more data. Create a new C project in Visual Studio called CH9-StringHeap. Enter the following code:

```
#include <stdio.h>
#include <stdlib.h>
#include <string.h>

void main() {

        char* cptr = (char*)malloc(40);
        char* cptrstart = cptr;
        char* cptr2;
        if (cptr == NULL) {

                printf("Failed to allocate memory\n");
                exit(1);
        }

        strcpy_s(cptr, sizeof(char)*40, "Working with Heap memory is
fun!");
        printf("%s\n", cptr);


        cptr2 = (char*)realloc(cptr, 80);
        cptrstart = cptr2;
        if (cptr2 == NULL) {
                printf("Failed to allocate more memory\n");
                exit(1);
        }

        strcat_s(cptr2, sizeof(char) * 80, " Make sure you track the Heap
size.");
        printf("%s\n", cptr2);

        free(cptrstart);

        for (;;);
}
```

Run the program and you should see the following output:

Working with Heap memory is fun!
Working with Heap memory is fun! Make sure you track the Heap size.

The program allocates 80 bytes for characters. The strpy() function copies a string into the Heap. The contents are then printed to the string. The realloc() function is called to increase the Heap size. The realloc() function creates a new Heap space with the larger size, and the address is assigned to a new pointer. The contents from the previous memory allocation are moved to the new memory allocation, and the pointer points to the first character in the Heap. The contents of the old Heap are moved to the new expanded Heap. There is no need to do a free(cptr) since the realloc() function frees the memory after the contents have been moved to the new block of memory. The new pointer is used from here on out to access the Heap. The string in memory is concatenated with a new string. The whole string is then printed out using cptr2. The cptrstart pointer is first set to the address of cptr, later it is reassigned to the new cptr2 address, and then it is used to free the memory.

The actions of memory allocation, freeing, and reallocation at the top level seem very straightforward. Under the covers, however, is a fair amount of complexity to implement a smooth functioning system. During the initial allocations, before any allocations are freed, the operation is simple. A pointer is kept that gives the address of the start of the free memory pool along with the size of available free memory. When a block of memory is requested, the allocation routine checks that the amount of memory requested is equal-to or less-than the available free memory; and if that is true, returns the pointer to the base of the free memory. Then the allocation routine changes the pointer to free memory to point to the free memory after this allocation and adjusts the amount of free memory available to reduce it by the amount just allocated. Now, after many allocations, what happens when a block of allocated memory is freed? That block of memory could be anywhere in the memory space of allocated memory. It may not be contiguous to the block of remaining free memory. So, the free routine has to now create a list of pointers to available blocks of memory along with the size of each of these memory blocks. The next time that a memory allocation is done, the allocation routine must look at the list of pointers to free memory, and find one that has at least the amount of memory being requested. It can then return that pointer, but it now must manage the list of free memory pointers to account for the block of memory allocated. If the free block is just the right size, its pointer is just removed from the list. If the free block is larger than the block allocated, the pointer must be adjusted for the memory allocated, and the size of the free block reduced by the amount of memory allocated. Imagine if this is all that the allocation and free routines do. If a program does a lot of dynamic memory allocation and freeing of blocks of memory of differing sizes, after a while, what started out to be a large block of memory available for use, starts to look like Swiss cheese with blocks of allocated and unallocated memory of varying size spread throughout that original pool of memory. If this kind of memory allocation and freeing goes on for some time, you might reach a point where an allocation request is made, the total of all blocks of free memory is more than enough to satisfy the request, but there is no one block of contiguous memory that can satisfy the request. The memory allocation request fails even though there is enough free memory, but not enough contiguous free memory. To deal with this situation, the memory allocation and free routines have to have the ability to essentially move blocks of memory around to take smaller blocks of memory and put them together in a larger, contiguous block. This is known as garbage collection and memory compaction. This is some of the magic of memory management that goes on under the covers as a program is running. Garbage collection or compaction could be done each time a block of memory is freed, or the memory management routine could monitor the fragmentation of the memory pool, and do garbage collection or compaction when the memory is becoming too fragmented. Both methods have pluses and minuses. If you do garbage collection with every allocation and free cycle, that slows the process down each time. If you only do garbage collection when the pool becomes too fragmented, the general allocation/free processing will be faster, but each time garbage collection or compaction is fired off due to fragmentation, the system will slow down until the defragmentation is complete. One other thing that is particularly important to keep track of is memory that is allocated and then is no longer going to be used. It should always be freed. If you write code that is so unstructured that there are cases when unused memory is occasionally not freed, this memory becomes permanently unavailable to the program for future use. If this happens enough and the program runs long enough, you might reach a condition where there is no memory available to be allocated. Most of the memory is lost to the memory allocations that were not properly freed. This condition is known as memory leaks, so design your programs well and structure them cleanly so that all allocated memory that is no longer needed is freed when it is no longer in use.

9.2.4 Computer Activity 9.4: Structures

Let's return to the mail address structure and see how to allocate an array of structures in Heap memory. Create a new C project in Visual Studio called CH9-AddressBook. Enter the following code:

```c
#include <stdio.h>
#include <stdlib.h>
#include <string.h>

typedef struct {
        char contactName[20];
        char contactAddress[30];
        char contactCityStZip[50];
        char contactPhone[16];
        char contactEmail[30];
}Contacts;


void main() {


        Contacts* mycontacts = (Contacts*)calloc(1000,sizeof(Contacts));
        Contacts* mycontactsStart = mycontacts;

        if (mycontacts == NULL) {
                printf("Failed to allocate memory");
                exit(1);
        }

        strcpy_s(mycontacts->contactName, 20, "Beth");
        strcpy_s(mycontacts->contactAddress, 30, "1 Smith St.");
        strcpy_s(mycontacts->contactCityStZip, 50, "Small Town, CT,
02030");
        strcpy_s(mycontacts->contactPhone, 16, "555-555-5555");
        strcpy_s(mycontacts->contactEmail, 30, "beth@clang.com");

        mycontacts++;

        strcpy_s(mycontacts->contactName, 20, "Ralph");
        strcpy_s(mycontacts->contactAddress, 30, "6 Main St.");
        strcpy_s(mycontacts->contactCityStZip, 50, "Ghost, CA, 92134");
        strcpy_s(mycontacts->contactPhone, 16, "888-555-5555");
        strcpy_s(mycontacts->contactEmail, 30, "ralph@clang.com");

        mycontacts++;

        strcpy_s(mycontacts->contactName, 20, "Jill");
```

```
        strcpy_s(mycontacts->contactAddress, 30, "1234 Cedar St.");
        strcpy_s(mycontacts->contactCityStZip, 50, "Big Bend, MT, 59030");
        strcpy_s(mycontacts->contactPhone, 16, "777-777-5555");
        strcpy_s(mycontacts->contactEmail, 30, "Jill@clang.com");

        mycontacts = mycontactsStart;

        printf("%s\n", mycontacts->contactName);
        printf("%s\n", mycontacts->contactAddress);
        printf("%s\n", mycontacts->contactCityStZip);
        printf("%s\n", mycontacts->contactPhone);
        printf("%s\n", mycontacts->contactEmail);

        printf("\n");
        mycontacts++;
        mycontacts++;

        printf("%s\n", mycontacts->contactName);
        printf("%s\n", mycontacts->contactAddress);
        printf("%s\n", mycontacts->contactCityStZip);
        printf("%s\n", mycontacts->contactPhone);
        printf("%s\n", mycontacts->contactEmail);

        free(mycontactsStart);

        for (;;);
}
```

Run the program and the contact information for Beth and Jill will appear on the console output. The program first defined a new structure type called contacts. The structure contains different character strings of different lengths for each contact item. The calloc() function is called to allocate 1000 of the Contact structures in memory and set all items to 0. We can see the 0s filled in when viewing memory in Visual Studio.

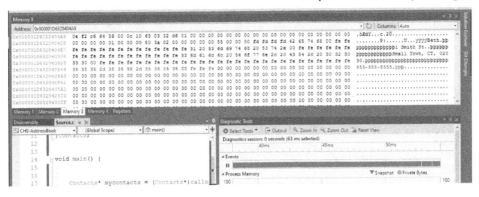

Three contacts are added to the allocated memory, and the first and last contacts are printed out. Memory is freed at the end. Unlike trying to use Stack space, Visual Studio didn't complain about the memory requirements since we allocated Heap space to store the 1000 contacts. The program is a demonstration of using a Heap to store an array of a structure. A real address book program would use a file to store the data, so that it is not lost when the program closes. We have also used sizeof(Contacts) for the size of the structure because we cannot be sure how the compiler will pack the elements of the struct into memory. Simply adding up the sizes of each element of the structure may give a size different from what the compiler actually allocated.

9.3 *Summary*

Unlike managed code languages like C# and Java, C forces the programmer to manage memory. It is not as scary as one would think. The key is understanding the Stack versus Heap trade-offs. If a data structure is going to be larger than the stack size, then the data needs to be moved to Heap. Only pointers are used when accessing Heap memory so careful planning and tracking of each pointer allocated is a requirement. The most important thing is to free the memory once it is no longer needed and before the program exits.

This ends the basic coverage of the C language and using Visual Studio to create programs based on C. There are more advanced topics to explore such as networking and creating a link list in C. The purpose of this book is to get you familiar with C with so much attention being given to programming MCUs, FPGAs, and applications that connect to Azure. The last two chapters introduce the Azure IoT C SDK and Azure Sphere.

10 Introduction to Azure IoT C SDK

The Internet of Things (IoT) is big a driving force for cloud computing. With billions of devices to be connected to the internet, there is a big reason why Microsoft is making the investment to support different ARM MCUs and different RTOSes connecting to Azure. Their first effort was to develop different IoT Device SDKs for Windows and Linux so developers could write applications that can connect to Azure. One of these SDKs is the Azure IoT C SDK, which is what we will be covering in this chapter.

10.1 Different Azure IoT SDKs

In order to connect different devices to the Cloud, Microsoft developed the Azure IoT Hub as a way to interconnect devices and cloud services. As a network hub/router connects computers on a network, the same goes for IoT Hub. Of course, security is an important part of the connections, so each device gets its own connection string. The communication protocol from the device to Azure IoT Hub is the MQ Telemetry Transport (MQTT), which is a lightweight network protocol that was designed for resource-constrained devices like MCUs. The Azure IoT Hub also supports HTTP and AMQP protocols, but MQTT has the most support for IoT. With MQTT, there is a server that acts as a broker to handle events and messages from all the devices connected to it. The Azure IoT Hub acts as a quasi-MQTT message broker that can route messages to the appropriate destination. The reason for the qualifier "quasi" is that according to Microsoft, the Azure IoT Hub is not a full MQTT broker.

One could attempt to use MQTT directly in an application, but the limits of the Azure IoT Hub have to be addressed. To help developers write applications to connect to the Azure IoT Hub, different IoT Device SDKs have been developed to address different programming languages and device types. The online documentation from Microsoft attempts to explain the different SDKs, but the terminology and the key introductory information are confusing and scattered throughout the site. Microsoft is addressing two types of devices:

- General-purpose devices like PCs, tablets, and smartphones that run Windows, iOS, Android, or Linux. By definition, general-purpose devices have plenty of RAM and disk space to run many applications.
- Embedded devices have limited memory and other resources. For example, a thermostat, HMI, engine controller, IP camera, etc. By definition, an embedded device will only run a single or limited number of applications.

For general-purpose devices, Microsoft has developed IoT SDKs for different programming languages: C# (.NET), Python, Node.js, Java, and C. It is the Azure IoT C SDK that we are introducing in this chapter.

For the Embedded devices that run on MCUs, C is the most popular programming language. The Azure IoT C SDK was designed for general-purpose devices, which have lots of memory and processing power compared to the smaller MCUs. To address resource-constrained devices, Microsoft created a derivative of the Azure IoT C SDK called the Embedded C SDK. The Embedded C SDK is for systems running Eclipse ThreadX , FreeRTOS, or no RTOS (aka Bare Metal). Since the RTOSes and bare metal solutions require a deeper discussion, the Embedded C SDK will be saved for future books and articles.

10.2 Computer Activity 10.1: Send a Message to Azure IoT Hub

In addition to the SDKs, there are a number of sample applications and hardware platforms to get familiar with to program for Azure IoT. We will take one of the examples, and rebuild it in Visual Studio to connect to Azure IoT Hub. This computer activity is based on the telemetry sample application where a message is sent 5 times to Azure IoT Hub. To test the example, there are a number of steps from downloading the tools and packages from Git Hub, to setting up Azure IoT Hub, and finally, writing the application.

10.2.1 Download and Install Support Tools

The first step is to download some tools to help with development. The first tool is the Git utility, so we can download the SDK and other tools from Git Hub.

1. Open a Web Browser.
2. Go to https://git-scm.com/downloads and download the Git installer for Windows.
3. Run the installer. Once installed, the Git utility can be run from the command prompt.

The second utility is CMake, which is used to build the code projects from the command line.

1. Open a Web Browser.
2. Go to https://cmake.org/download and download the CMake installer for Windows.
3. Run the installer. Once installed, the CMake utility can be run from the command prompt.

The final utility is the Azure CLI (command line interface). We will need this tool to interact with Azure and the IoT Hub.

1. Open a Web Browser.
2. Go to Install the Azure CLI for Windows | Microsoft Learn and download the Azure CLI installer for Windows.
3. Run the installer. Once installed, the az.exe utility can be run from the command prompt.

10.2.2 Azure Portal – Create the Azure IoT Hub and Device

With the utilities installed, we next turn to creating an Azure IoT Hub and a device connected to the Azure IoT Hub. To do this, you will need to create an Azure account.

Note: Microsoft is constantly updating Azure. The steps here might change over time. Also, you can delete the resource created for this computer activity after completion.

1. Open a browser.
2. Go to https://azure.microsoft.com/en-us and register for an account. You will have to provide a payment option and select a subscription type.

3. Log into your Azure Portal account: https://portal.azure.com.
4. In the portal, click on the 3 bars to the top left to open the side menu.
5. Click the "+ Create a resource" menu item.

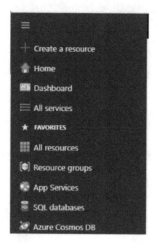

6. Click on the Internet of Things.

Home >

Create a resource …

Get Started

Recently created

Categories

AI + Machine Learning

Analytics

Blockchain

Compute

Containers

Databases

Developer Tools

DevOps

Identity

Integration

Internet of Things

IT & Management Tools

7. Click Create under IoT Hub

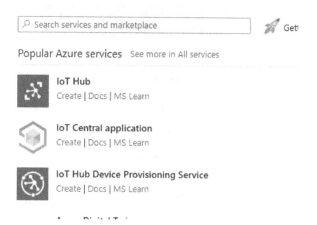

8. The first screen provides the basic settings to select the Subscription, resource group (which you will have to create), the name of the IoT Hub, the tier, and the Daily message limit. Fill in the information as you see fit for your development.

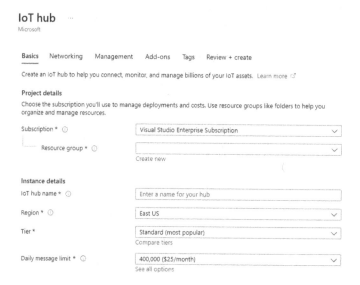

9. Click on Networking – Here you can set the connectivity configuration and Minium TLS version. For this computer activity, you can leave the defaults.
10. We can skip the other options, click on "Review + Create" and finish the steps to create the IoT Hub.
11. Click on the newly created IoT Hub.

12. In the menu on the left, click on Devices

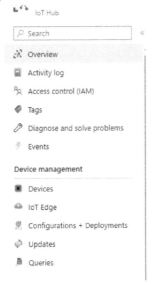

13. Click on the + Add Device

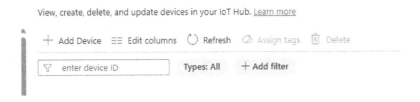

14. A window appears. Add the Device ID name of the device in lowercase. For this example, we will use "winmessagesample".
15. Click Save.
16. Go back to the Azure IoT Hub and Devices, and you should see your device listed in the table.

Device ID	Type	Status	Last status update	Authentication type
winmessagesample	IoT Device	Enabled	--	Shared Access Signature

17. Click on the device ID and a screen will appear with the connection and keys for use in our application.

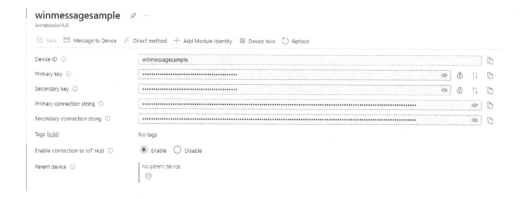

18. Click on the copy symbol 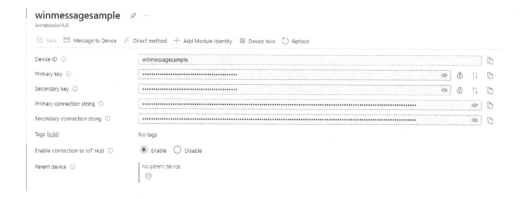 next to the "Primary Connection string".
19. Open Notepad, paste the string into Notepad, and save the file. We will need this connection string for your program.

10.2.3 Downloads from Git HUB and Build the Library for Use with Visual Studio

We want to download the full Azure IoT SDK C package and the Azure IoT SDK C vcpkg.

1. Open File Explorer and create a folder called Azure-sdk-C.
2. Open PowerShell.
3. Change directory to the \Azure-sdk-c folder.
4. Run the following command to download the vcpkg to a new folder:

   git clone https://github.com/Microsoft/vcpkg.git vcpkg_new

The output will look similar to this:

```
PS  E:\azure-sdk-c>  git  clone  https://github.com/Microsoft/vcpkg.git
vcpkg_new
Cloning into 'vcpkg_new'...
remote: Enumerating objects: 198160, done.
remote: Counting objects: 100% (29183/29183), done.
```

```
remote: Compressing objects: 100% (2674/2674), done.
remote: Total 198160 (delta 27128), reused 26598 (delta 26509), pack-reused
168977Receiving objects: 100% (198160/198160), 51.20 MiB | 14.60 MiB/s
Receiving objects: 100% (198160/198160), 62.15 MiB | 15.55 MiB/s, done.
Resolving deltas: 100% (130843/130843), done.
Updating files: 100% (10318/10318), done.
```

5. Once the download has completed, change directory to the new \Azure-sdk-C\ vcpkg_new folder.

6. Run the following:

    .\bootstrap-vcpkg.bat

The output will look similar to this:

```
PS E:\azure-sdk-c\vcpkg_new> .\bootstrap-vcpkg.bat
Downloading                              https://github.com/microsoft/vcpkg-
tool/releases/download/2023-06-22/vcpkg.exe        ->        E:\azure-sdk-
c\vcpkg_new\vcpkg.exe... done.
Validating signature... done.

Telemetry
---------
vcpkg collects usage data in order to help us improve your experience.
The data collected by Microsoft is anonymous.
You can opt-out of telemetry by re-running the bootstrap-vcpkg script with
-disableMetrics,
passing --disable-metrics to vcpkg on the command line,
or by setting the VCPKG_DISABLE_METRICS environment variable.

Read more about vcpkg telemetry at docs/about/privacy.md
```

7. Now we need to create the library files that are needed for Visual Studio. Run the following:

    .\vcpkg.exe integrate install

The output will look similar to this:

```
PS E:\azure-sdk-c\vcpkg_new> .\vcpkg.exe integrate install
Applied user-wide integration for this vcpkg root.
```

CMake projects should use: `"-DCMAKE_TOOLCHAIN_FILE=E:/azure-sdk-c/vcpkg_new/scripts/buildsystems/vcpkg.cmake"`

All MSBuild C++ projects can now #include any installed libraries. Linking will be handled automatically. Installing new libraries will make them instantly available.

8. Build the library for X64 Windows:

    .\vcpkg.exe install azure-iot-sdk-c:x64-windows

The output is a bit lengthy. Several folders get created in the \Azure-sdk-C\ vcpkg_new folder:

\buildtress
\downloads
\packages
\installed

The \Azure-sdk-C \vcpkg_new\installed\x64-windows\include contains the header files and libraries for use with application development.

9. Finally, download the full Azure IoT C SDK. Change directory to \Azure-IoT-C
10. Run the following:

    git clone https://github.com/Azure/azure-iot-sdk-c.git

The \Azure-IoT-C \azure-iot-sdk-c-main\iothub_client\samples contains several sample applications to learn from. We will rebuild the iothub_ll_telemetry_sample in Visual Studio.

10.2.4 Message Application

The sample applications that come with the SDK have .sln files, so you can quickly open the projects in Visual Studio. By just using the supplied .sln file, you miss out on the little details needed to create the solution. Creating a project from scratch is best to learn all the little steps to set up and build a project. Please be aware that this is a very simple, not fully secure application. It is important to address the best practices called out in the online documentation to avoid any denial of service or security issues.

1. Open Visual Studio 2022.
2. Select "Create a new project".
3. Select the Basic C template from the template list, and click Next.
4. Enter the name IoTHUBMessage, select the directory, and click Create.
5. The project needs to point to the Azure IoT C SDK include files. From the menu select Project->IoTHUBMessage properties.
6. Set the Configuration to All Configurations
7. Under Configuration Properties go to C/C++-> General.
8. In the Addition Include Directories, add the path (\Azure-sdk-C \vcpkg_new\installed\x64-windows\include) to the include files generated in section 11.2.3.

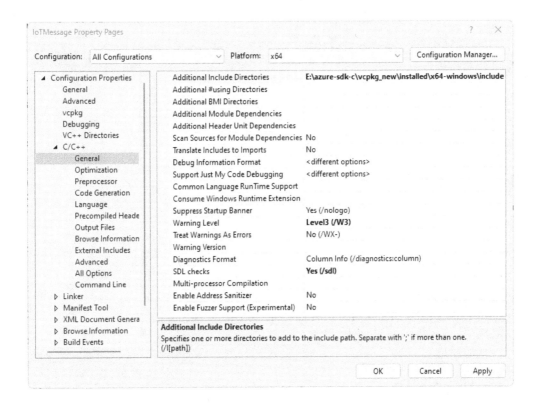

9. Next, go to Linker->Input.

10. In the Additional Dependencies add the following files:

rpcrt4.lib
ncrypt.lib
crypt32.lib
winhttp.lib
secur32.lib
ws2_32.lib

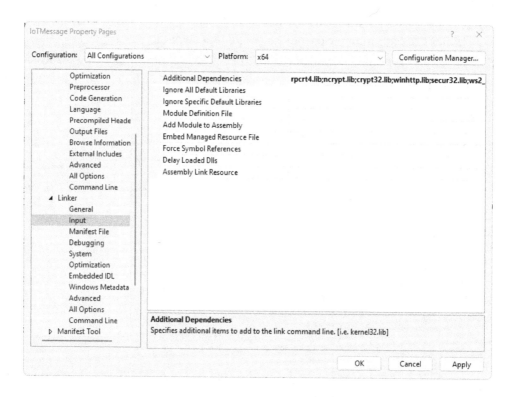

These extra libraries are needed for encryption and network communication.

11. Click Apply.
12. Click Ok.
13. Open Source.c.
14. Add the following includes before the main() function:

```
#include <stdio.h>
#include <stdlib.h>

#include "iothub.h"
#include "iothub_device_client_ll.h"
#include "iothub_client_options.h"
#include "iothub_message.h"
#include "azure_c_shared_utility/threadapi.h"
#include "azure_c_shared_utility/crt_abstractions.h"
#include "azure_c_shared_utility/shared_util_options.h"
#include "iothubtransportmqtt.h"
```

15. After the includes and before main(), add the following constants:

```
static const char* connectionString = "<enter connection string that
was saved in section 11.2.2>";
#define MESSAGE_COUNT           5
static bool g_continueRunning = true;
static size_t g_message_count_send_confirmations = 0;
```

16. Next, two callbacks are needed to address the Azure IoT Hub connection and message handling. After the constants and before main() enter the following:

```
static void send_confirm_callback(IOTHUB_CLIENT_CONFIRMATION_RESULT
result, void* userContextCallback)
{
    (void)userContextCallback;
    // When a message is sent this callback will get invoked
    g_message_count_send_confirmations++;
    (void)printf("Confirmation callback received for message %lu with
result %s\r\n", (unsigned long)g_message_count_send_confirmations,
MU_ENUM_TO_STRING(IOTHUB_CLIENT_CONFIRMATION_RESULT, result));
}

static void
connection_status_callback(IOTHUB_CLIENT_CONNECTION_STATUS result,
IOTHUB_CLIENT_CONNECTION_STATUS_REASON reason, void* user_context)
{
    (void)reason;
    (void)user_context;
    // This sample DOES NOT take into consideration network outages.
    if (result == IOTHUB_CLIENT_CONNECTION_AUTHENTICATED)
    {
        (void)printf("The device client is connected to iothub\r\n");
    }
    else
```

```
    {
        (void)printf("The device client has been disconnected\r\n");
    }
}
```

17. Finally, enter the following in main():

```
int main(void) {

    IOTHUB_MESSAGE_HANDLE message_handle;
    size_t messages_sent = 0;
    const char* telemetry_msg = "Hello from Windows Device";

    // Used to initialize IoTHub SDK subsystem
    (void)IoTHub_Init();

    IOTHUB_DEVICE_CLIENT_LL_HANDLE device_ll_handle;

    (void)printf("Creating IoTHub Device handle\r\n");
    // Create the iothub handle here
    device_ll_handle =
IoTHubDeviceClient_LL_CreateFromConnectionString(connectionString,
MQTT_Protocol);
    if (device_ll_handle == NULL)
    {
        (void)printf("Failure creating IotHub device. Hint: Check
your connection string.\r\n");
    }
    else
    {
        bool traceOn = true;
        IoTHubDeviceClient_LL_SetOption(device_ll_handle,
OPTION_LOG_TRACE, &traceOn);

        // Setting connection status callback to get indication of
connection to iothub

(void)IoTHubDeviceClient_LL_SetConnectionStatusCallback(device_ll_han
dle, connection_status_callback, NULL);

        do
        {
            if (messages_sent < MESSAGE_COUNT)
            {
                // Construct the iothub message from a string or a
byte array
```

```
                message_handle =
IoTHubMessage_CreateFromString(telemetry_msg);

                // Add custom properties to message
                (void)IoTHubMessage_SetProperty(message_handle,
"property_key", "property_value");

                (void)printf("Sending message %d to IoTHub\r\n",
(int)(messages_sent + 1));

IoTHubDeviceClient_LL_SendEventAsync(device_ll_handle,
message_handle, send_confirm_callback, NULL);

                // The message is copied to the sdk so the we can
destroy it
                IoTHubMessage_Destroy(message_handle);

                messages_sent++;
            }
            else if (g_message_count_send_confirmations >=
MESSAGE_COUNT)
            {
                // After all messages are all received stop running
                g_continueRunning = false;
            }

            IoTHubDeviceClient_LL_DoWork(device_ll_handle);
            ThreadAPI_Sleep(1);

        } while (g_continueRunning);

        // Clean up the iothub sdk handle
        IoTHubDeviceClient_LL_Destroy(device_ll_handle);
    }
    // Free all the sdk subsystem
    IoTHub_Deinit();

    printf("Press any key to continue");
    (void)getchar();

    return 0;
}
```

18. Save the file.

19. Build the file and make sure that there are no errors.

20. Open PowerShell or command window and run the az.exe utility to log into your Azure account:

az.exe login

If you have multiple tenants/domains in your account, you will have to specify the tenant you want to access:

az.exe login --allow-no-subscriptions --tenant <tenant domain name>.com

You can find all the tenants by clicking on the ⚙ symbol in Azure Portal.

21. Before we run the program, we need to run an AZ CLI command to capture messages/events coming into IoT Hub. Run the following command with the correct Hub name and device ID:

Az.exe iot hub monitor-events --hub-name HubName --device-id DeviceID

22. In Visual Studio, run the debugger. The program will start a command window session and output the following:

```
Creating IoTHub Device handle
Sending message 1 to IoTHub
Sending message 2 to IoTHub
Sending message 3 to IoTHub
Sending message 4 to IoTHub
Sending message 5 to IoTHub
-> 21:03:03 CONNECT | VER: 4 | KEEPALIVE: 240 | FLAGS: 192 | USERNAME:
customHUB.azure-devices.net/WinSampleMessage/?api-version=2020-09-
30&DeviceClientType=iothubclient%2f1.10.0%20(native%3b%20WindowsProduct%3
a0x00000030%206.2%3b%20x64%3b%20%7b172FE9F7-FB03-47B1-8955-
E1B217C3377C%7d) | PWD: XXXX | CLEAN: 0
<- 21:03:03 CONNACK | SESSION_PRESENT: false | RETURN_CODE: 0x0
The device client is connected to iothub
-> 21:03:03 PUBLISH | IS_DUP: false | RETAIN: 0 | QOS: DELIVER_AT_LEAST_ONCE
|                                                      TOPIC_NAME:
devices/WinSampleMessage/messages/events/property_key=property_value     |
PACKET_ID: 2 | PAYLOAD_LEN: 12
-> 21:03:03 PUBLISH | IS_DUP: false | RETAIN: 0 | QOS: DELIVER_AT_LEAST_ONCE
|                                                      TOPIC_NAME:
devices/WinSampleMessage/messages/events/property_key=property_value     |
PACKET_ID: 3 | PAYLOAD_LEN: 12
```

```
-> 21:03:03 PUBLISH | IS_DUP: false | RETAIN: 0 | QOS: DELIVER_AT_LEAST_ONCE
|                                                           TOPIC_NAME:
devices/WinSampleMessage/messages/events/property_key=property_value      |
PACKET_ID: 4 | PAYLOAD_LEN: 12
-> 21:03:03 PUBLISH | IS_DUP: false | RETAIN: 0 | QOS: DELIVER_AT_LEAST_ONCE
|                                                           TOPIC_NAME:
devices/WinSampleMessage/messages/events/property_key=property_value      |
PACKET_ID: 5 | PAYLOAD_LEN: 12
-> 21:03:03 PUBLISH | IS_DUP: false | RETAIN: 0 | QOS: DELIVER_AT_LEAST_ONCE
|                                                           TOPIC_NAME:
devices/WinSampleMessage/messages/events/property_key=property_value      |
PACKET_ID: 6 | PAYLOAD_LEN: 12
<- 21:03:04 PUBACK | PACKET_ID: 2
Confirmation    callback    received    for    message    1    with    result
IOTHUB_CLIENT_CONFIRMATION_OK
<- 21:03:04 PUBACK | PACKET_ID: 3
Confirmation    callback    received    for    message    2    with    result
IOTHUB_CLIENT_CONFIRMATION_OK
<- 21:03:04 PUBACK | PACKET_ID: 4
Confirmation    callback    received    for    message    3    with    result
IOTHUB_CLIENT_CONFIRMATION_OK
<- 21:03:04 PUBACK | PACKET_ID: 5
Confirmation    callback    received    for    message    4    with    result
IOTHUB_CLIENT_CONFIRMATION_OK
<- 21:03:04 PUBACK | PACKET_ID: 6
Confirmation    callback    received    for    message    5    with    result
IOTHUB_CLIENT_CONFIRMATION_OK
-> 21:03:04 DISCONNECT
Press any key to continue
```

In the command window running the az CLI hub monitor, you will see the arriving message:

```
Dependency update (uamqp 1.2) required for IoT extension version: 0.21.3.
Continue? (y/n) -> y
Updating required dependency...
Update complete. Executing command...
Starting event monitor, filtering on device: WinSampleMessage, use ctrl-c
to stop...
event:
  component: ''
  interface: ''
  module: ''
  origin: WinSampleMessage
  payload: test_message
```

```
event:
  component: ''
  interface: ''
  module: ''
  origin: WinSampleMessage
  payload: test_message

event:
  component: ''
  interface: ''
  module: ''
  origin: WinSampleMessage
  payload: test_message

event:
  component: ''
  interface: ''
  module: ''
  origin: WinSampleMessage
  payload: test_message

event:
  component: ''
  interface: ''
  module: ''
  origin: WinSampleMessage
  payload: test_message
```

23. Close the application debug window by hitting enter twice. The action will stop the debugger.
24. In the console window running the az CLI hub monitor, log out of the az CLI:

    az.exe logout

25. Close the command window

10.3 Summary

For most general-purpose systems, you will most likely use the other SDKs to create GUI feature-rich applications. The ability to write an Azure IoT client application in C allows you to create utilities to interact with your Azure IoT Hub, devices, and data in Azure. In this chapter, we explored how to set up the Azure IoT C SDK and attach the libraries to a C project we created from scratch in Visual Studio. So far, we have written all our applications in Windows, In the next chapter we will use Visual Studio to write an application for an MCU.

11 Writing C Applications for Azure Sphere

As stated in Chapter 1, the goal of this book was to provide an introduction to the C language as a lead-in to future books and articles that will focus on programming MCUs, FPGAs, and connecting to Azure. Microsoft is making investments to support the billions of devices that will be connecting to the Cloud. The previous chapter introduced writing applications that can talk to the Cloud using the Azure IoT C SDK, and it discussed how the Azure IoT C SDK has been expanded to support other MCUs and RTOSes. There is one remaining piece to Microsoft's focus on MCUs and that is their own custom MCU called Azure Sphere. The goal of this chapter is to get the Azure Sphere development environment set up and test a couple of C applications.

11.1 Azure Sphere Introduction

Microsoft researchers looked at the issues for connected devices and released a paper titled: *The Seven Properties of Highly Connected Devices.* The research led to the creation of a custom silicon solution with a security subsystem built in that implements certificate security from hardware to application. The final result is a technology called Azure Sphere. The MediaTek MT3620 is the first silicon to implement the solution to be marketed as Azure Sphere, and there will be other silicon vendors in the future offering solutions. Azure Sphere is a complete system-on-chip solution that includes a multi-core ARM solution (ARM Cortex-A7 and ARM Cortex-M4) with all the popular MCU I/O capabilities such as UART, I2C, SPI, PWM, ADC, and GPIO. That is not all. The chip also includes RAM and flash storage memory that hosts a tiny Linux kernel running on one of the ARM cores to handle all the Azure connections. Instead of device manufacturers having to maintain an operating system, Azure Sphere reduces the development effort to simply writing a C-language application to access the I/O and send data to Azure. Maintenance for the Linux kernel is handled by Microsoft and the OEM can control when the Linux kernel gets updated.

11.2 Avnet Azure Sphere MT3620 Starter Kit

There have been a few development boards based on the MT3620. The Avnet Azure Sphere MT3620 Starter Kit will be used for the sample applications in this chapter.

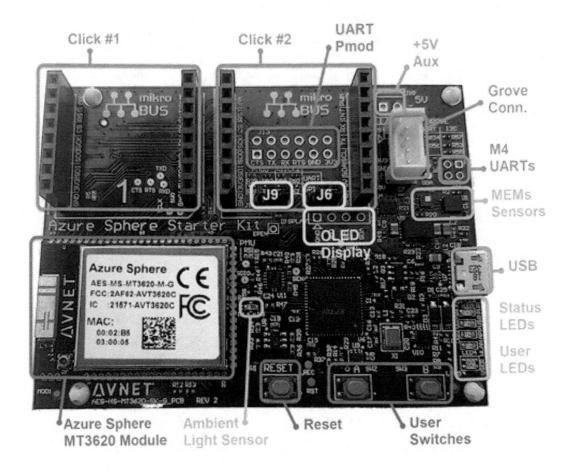

The Avnet Azure Sphere MT3620 Starter Kit provides all the basic MCU I/O: GPIO, SPI, PWM, ADC, and I2C, and it also includes peripherals for temperature, pressure, ambient light, and an accelerometer. Two microBUS connectors round out the highlights of the board.

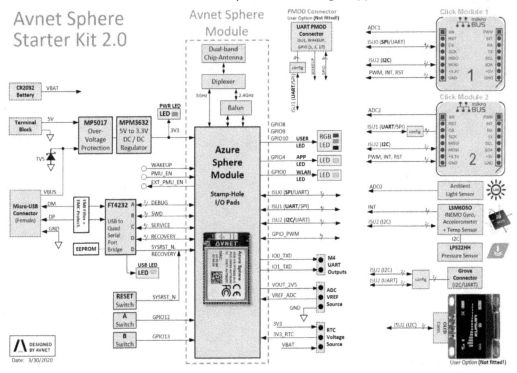

Although the computer activities are for the Avnet Azure Sphere MT3620 Starter Kit, the steps can be adapted to other Azure Sphere kits.

11.3 Computer Activity 11.1: Set Up the Azure Sphere Development Environment

For this computer activity, you will need the Avnet Azure Sphere MT3620 Starter Kit (or equivalent platform) and a Wi-Fi connection to the Internet. The setup is broken into two parts: software setup and device setup.

11.3.1 Part 1: Set Up the SDK and Sample Applications for Visual Studio

If you are reading this chapter, you should already have Visual Studio 2022 installed on your development system.

Note: This section refers to links that can change over time, so you may have to do an Internet search to find some of the items if they have been moved.

1. Created a folder on your development machine c:\AzureSphere. This will be the location for downloads and applications.
2. Click on this link to download the Azure Sphere SDK.
3. After the download has been completed, run the Azure_Sphere_SDK_xxxx.exe installer.

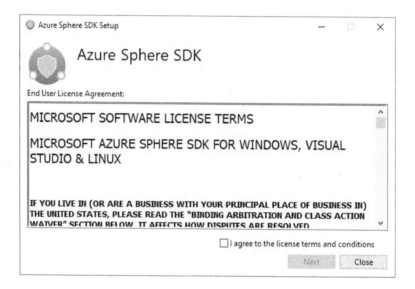

4. Walk through the install wizard to install the SDK. The SDK installs the Azure Sphere CLI.
5. Click on this link Visual Studio Extension for Azure Sphere, and download the Azure Sphere for Visual Studio 2022 vsix file.
6. Double-click on the Azure_Sphere_VS2022_22_07_RC2.vsix to run the installer. You may have to re-run the Visual Studio installer to address any prerequisites to support Linux development.

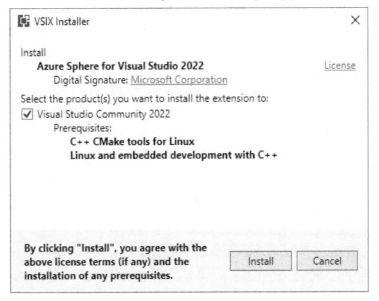

7. Click on Install and click Close when finished. The Azure_Sphere_VS2022_22_07_RC2.vsix adds the Azure Sphere project templates to Visual Studio.
8. The final item is to download the Azure Sphere Samples from GitHub: https://github.com/Azure/azure-sphere-samples.
9. Click on the Code button and select Download Zip.
10. Put the zip file in c:\AzureSphere and extract the contents. We will use some of the contents in the upcoming computer activities.

11.3.2 Part 2: Device Setup

With the software setup complete, we can now set up the hardware. The Azsphere CLI provides a complete interface to manage the Azure Sphere devices. The Azsphere CLI will be used to set up the device.

1. Open Device Manager.
2. Connect the Azure Sphere Kit to your development machine via the USB cable.
3. The USB serial drivers should install for the kit. If there are problems, please contact your manufacturer.
4. Open PowerShell with administrative privileges.

5. If you have never created an Azure Sphere account, you will need to run the following:

```
azsphere register-user --new-user <email-address>
```

6. Follow the steps to create a new account with the password.
7. If you have already created an Azure Sphere account, you can run the following to login:

```
azsphere login
```

8. The next step is to create a tenant. The tenant is a security feature that helps limit those who can program the device. First, see if there are any tenants for the device. Run the following:

```
azsphere tenant list
```

9. A new kit will not have any tenants, so you will need to create one. Run the following and provide a name for the tenant:

```
azsphere tenant create --name <tenant-name or tenant-ID>
```

10. Run the following to get the tenant's name and tenant ID:

```
azsphere tenant list
```

11. Copy down the tenant's name and tenant ID and save the information to a text file.
12. Next, we need to claim the device. Once this has been performed, it is locked to the tenant. It can never be undone. Run the following:

```
azsphere device claim
```

13. Next, connect the device to a Wi-Fi network by running the following command

```
azsphere device wifi add --ssid "WiFi SSID" --psk "password key
with quotes"
```

or open Visual Studio. Select "Continue without code". From the menu, select View->Other Windows->Azure Sphere Explorer. From Azure Sphere Explorer, you can get a list of available networks. Locate your Wi-Fi, right-click, and select Connect. Follow the rest to connect to the Wi-Fi network.

14. Enable the device for application development:

```
azsphere device enable-development
```

We are ready to create a program.

11.4 Computer Activity 11.2: Blinking LED Application

Chapter 1 covered the universal getting started program: Hello World. For hardware development, the Hello World is blinking an LED. The computer activity will use a built-in template to create the program, but some modifications are needed to support the hardware.

1. Open Visual Studio.
2. Select Create new project.
3. In the All Platforms drop-down, select Azure Sphere. A list of available Azure Sphere project templates appears.

The first two templates are for building applications from scratch. We will discuss the differences between the two in the next computer activity. The Azure Sphere Blink is an almost ready-to-go solution.

4. Select "Azure Sphere Blink" and click Next.
5. Name the project "AzureSphereBlink1-Avnetkit2" and set the target folder for c:\AzureSphere.
6. Click Create. Visual Studio will perform some processing to create the project. Once completed, we need to make some modifications to support the hardware.
7. Open File Explorer and go to C:\AzureSphere\azure-sphere-samples-main\HardwareDefinitions
8. Right-click on avnet_mt3620_sk_rev2 and select copy.
9. Go to C:\AzureSphere\AzureSphereBlink1-Avnetkit2\AzureSphereBlink1-Avnetkit2\HardwareDefinitions and paste avnet_mt3620_sk_rev2 into the folder. The folder contains the hardware definitions for the kit.
10. In Visual Studio, open Solution Explorer for the project. Open CMakeList.txt.

11. Change the following line from this:

```
azsphere_target_hardware_definition(${PROJECT_NAME} TARGET_DIRECTORY
"HardwareDefinitions/mt3620_rdb" TARGET_DEFINITION
"template_appliance.json")
```

To this:

```
azsphere_target_hardware_definition(${PROJECT_NAME} TARGET_DIRECTORY
"HardwareDefinitions/avnet_mt3620_sk_rev2" TARGET_DEFINITION
"sample_appliance.json")
```

12. Save the file.
13. Open main.c.
14. A header file include statement needs to be set to the correct header file for the kit. Change the following line from this:

```
#include <hw/template_appliance.h>
```

to this:

```
#include <hw/sample_appliance.h>
```

15. Next, we need to change the LED name to match what is defined in the header file for the kit. Change the following line from this:

```
    int   fd   =   GPIO_OpenAsOutput(TEMPLATE_LED,   GPIO_OutputMode_PushPull,
GPIO_Value_High);
```

to this:

```
    int   fd   =   GPIO_OpenAsOutput(SAMPLE_LED,   GPIO_OutputMode_PushPull,
GPIO_Value_High);
```

16. Save the file.
17. Finally, open the app_manifest.json.
18. Change the following from this:

```
  "Gpio": [ "$TEMPLATE_LED" ],
```

To this:

```
  "Gpio": [ "$SAMPLE_LED" ],
```

19. Save the file.

20. Build the application, and you should see in the Output window that the build succeeded.

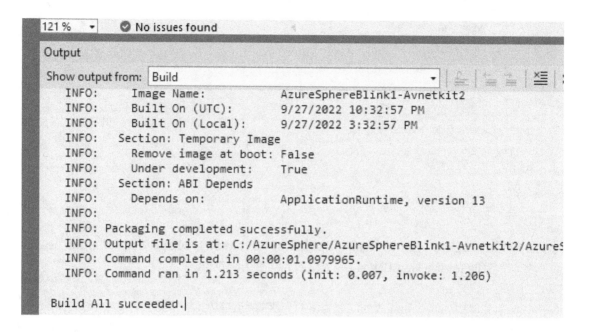

21. With the Avnet Azure Sphere MT3620 Starter Kit connected to the development machine, start the debugger.

The application will be downloaded to the device. Once completed, the Red LED will start blinking. The application is being remotely debugged just as we have been debugging local applications in the previous chapters.

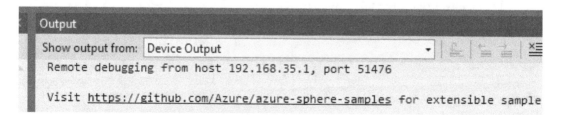

22. Stop the debug session when finished.

11.5 Azure Sphere Project Templates

There are two blank templates to create Azure Sphere applications: Azure Sphere HL Core Blank and Azure Sphere RTCore Blank. The Blink application we just built is based on the HL template. Azure Sphere is a multi-ARM core solution. One core is dedicated to running the internal Linux kernel that addresses the network and connects to Azure. The HL template targets the Linux kernel. The other ARM core is for running real-time applications. These real-time applications could be running in an RTOS like Eclipse ThreadX or raw on the processor. The Azure Sphere sample applications that were downloaded from GitHub contain different examples that demonstrate using both cores.

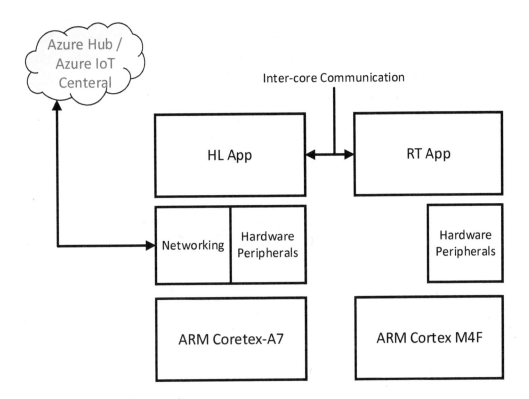

11.6 Computer Activity 11.3: Arrays and Pointers

In Chapter 1, the C standard libraries were highlighted. Different operating systems and hardware platforms will have additional libraries that support their development environments. In the last example, you will have noticed some familiar libraries being included and some new ones. The application was compiled using the GCC compiler which supports Linux. The applications before this were compiled with cl.exe compiler for Windows. The C language is based on a standard, so both compilers should support the same C standard libraries. Because of the C standard library, we can run the previous chapter projects on Azure Sphere with only some adjustments. In this computer activity, we will re-implement computer activity 5.4.

1. In Visual Studio, create a new Azure Sphere HL Core Blank application and call the application CH10-BasicArrayPointer.
2. We will not be accessing hardware, but it is a good habit to include the target hardware support files. Open File Explorer and go to C:\AzureSphere\azure-sphere-samples-main. Right-click on HardwareDefinitions and select copy. Yes, we will copy all the available hardware kits. You can later remove the hardware definitions that you are not using.
3. Go to C:\AzureSphere\CH10-BasicArrayPointer\CH10-BasicArrayPointer and paste HardwareDefinitions into the folder. The folder contains the hardware definitions for the kit.
4. Open CMakeList.txt and add the following line before the very last line in the file:

```
azsphere_target_hardware_definition(${PROJECT_NAME} TARGET_DIRECTORY
"HardwareDefinitions/avnet_mt3620_sk_rev2" TARGET_DEFINITION
"sample_appliance.json")
```

5. Save the file.
6. Open Main.c. There will be some code to send a message to the debug output. Fill in the rest of the code:

```
// This minimal Azure Sphere app prints "High Level Application" to the debug
// console and exits with status 0.

#include <applibs/log.h>
#include <stdio.h>

int main(void)
{
    // Please see the extensible samples at:
https://github.com/Azure/azure-sphere-samples
```

```
// for examples of Azure Sphere platform features
    Log_Debug("High Level Application\n");

        int scores[] = { 4,6,9,10,2,1 };

        int* scoreptr = scores;

        printf("scores address is %p\n", scores);
        printf("scoreptr is pointing to score address %p\n", scoreptr);
        printf("scoreptr is currently pointing to the value %d\n",
*scoreptr);

        scoreptr++;

        printf("scores[1] address is %p\n", &scores[1]);
        printf("scoreptr is pointing to score address %p\n", scoreptr);
        printf("scoreptr is currently pointing to the value %d\n",
*scoreptr);

        scoreptr += 2;

        printf("scores[3] address is %p\n", &scores[3]);
        printf("scoreptr is pointing to score address %p\n", scoreptr);
        printf("scoreptr is currently pointing to the value %d\n",
*scoreptr);

        for (;;); //Never gets to return.

    return 0;
}
```

7. Save the file.
8. Build the application. There should be no errors.
9. Set a breakpoint at the Int scores[] array.
10. Run the debugger.
11. Step through the code.

The standard debug output for the system is the Output window in Visual Studio. You can see the same messages as before.

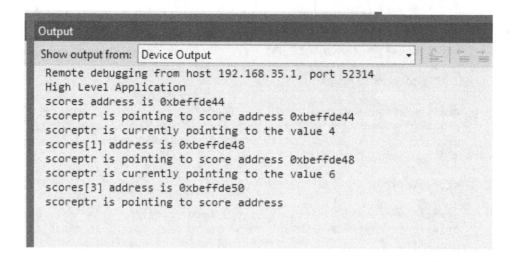

```
Output
Show output from:  Device Output
    Remote debugging from host 192.168.35.1, port 52314
    High Level Application
    scores address is 0xbeffde44
    scoreptr is pointing to score address 0xbeffde44
    scoreptr is currently pointing to the value 4
    scores[1] address is 0xbeffde48
    scoreptr is pointing to score address 0xbeffde48
    scoreptr is currently pointing to the value 6
    scores[3] address is 0xbeffde50
    scoreptr is pointing to score address
```

The debug support windows will be for the ARM processor and not X86. The disassembly window shows ARM assembly instructions.

```
Disassembly  ⊕ X  main.c      CMakeLists.txt      CMake Overview Pages
Address: main
  ⌄ Viewing Options
                   ...  ...  ...       ...
    0xbeeed658  add r3, pc
    0xbeeed65a  mov r0, r3
    0xbeeed65c  blx 0xbeeed4f0 <Log_Debug@plt>
●   0xbeeed660  ldr r3, [pc, #176]  ; (0xbeeed714 <main+212>)
    0xbeeed662  add r3, pc
    0xbeeed664  adds    r4, r7, #4
    0xbeeed666  mov r5, r3
    0xbeeed668  ldmia   r5!, {r0, r1, r2, r3}
    0xbeeed66a  stmia   r4!, {r0, r1, r2, r3}
    0xbeeed66c  ldmia.w r5, {r0, r1}
    0xbeeed670  stmia.w r4, {r0, r1}
    0xbeeed674  adds    r3, r7, #4
    0xbeeed676  str r3, [r7, #0]
    0xbeeed678  adds    r3, r7, #4
    0xbeeed67a  mov r1, r3
    0xbeeed67c  ldr r3, [pc, #152]  ; (0xbeeed718 <main+216>)
    0xbeeed67e  add r3, pc
    0xbeeed680  mov r0, r3
    0xbeeed682  blx 0xbeeed4c0 <printf@plt>
⇨   0xbeeed686  ldr r1, [r7, #0]
    0xbeeed688  ldr r3, [pc, #144]  ; (0xbeeed71c <main+220>)
    0xbeeed68a  add r3, pc
    0xbeeed68c  mov r0, r3
```

The registers window shows the ARM registers.

```
Registers                                                                                          ▾ 8 ×
  r0 = 0x1d r1 = 0xbeeed771 r2 = 0x0 r3 = 0xd2b8c400 r4 = 0xbeffde54 r5 = 0xbeeed818 r6 = 0xbeeed641  ▲
  r7 = 0xbeffde40 r8 = 0x20 r9 = 0x1 r10 = 0xbeffdea4 r11 = 0x0 r12 = 0xbeeed770 sp = 0xbeffde40
  lr = 0xbeeed687 pc = 0xbeeed686 cpsr = 0x60000030                                                  ▾
121 %  ▾
```

11.7 *Summary*

This concludes our short introduction to Azure Sphere. A more detailed book on Azure Sphere is required to cover all of the features that the hardware provides. Azure Sphere is a nice solution that allows you to focus solely on the application. All Linux kernel development is handled by Microsoft. There is only one Azure Sphere MCU, and it might not have enough horsepower to fit every application. The final chapter looks at Eclipse ThreadX which has support for many popular MCUs available on the market today.

12 Introduction to Eclipse ThreadX™

Microsoft's investment in the MCU market space did not end with Azure IoT C SDK and Azure Sphere. Microsoft acquired Express Logic, the developer of the ThreadX real-time operating system. ThreadX has support for many of the popular MCUs being used today. ThreadX was renamed to Azure RTOS; and as the name implies, Azure RTOS connects MCU devices to Azure. With a single acquisition, Microsoft has entrenched itself in the MCU market space and has enabled developers to connect MCUs to the Cloud. Then, just as we finished writing the first draft of this book, Microsoft gave Azure RTOS to the open-source community. The open-source version of this RTOS is now called Eclipse ThreadX. With what is known of this open-source version, we have attempted to update our references to the new Eclipse ThreadX naming, but not all references in the exercises could be updated at this time

Eclipse ThreadX is comprised of 6 components: the ThreadX™ kernel, NetX™ and NetXDuo™ for networking, FileX™ file system, USBX™ for a USB stack, TraceX™ for analysis, and GUIX™ for a 2D graphics solution. Since all the code has been written in C with some assembly, Azure RTOS has been ported to many different MCU architectures including a port that runs on Windows. As we were researching ThreadX, the fact that you can build and run a GuiX application in Windows using Visual Studio came to our attention. Therefore, the focus of this chapter is to demonstrate how to run a GuiX/ThreadX application on Windows.

12.1 GUIX and GUIX Studio
The theory of real-time operating systems and all the feature details of the ThreadX kernel are big topics to be discussed in other publications. Eclipse ThreadX is well documented on the product website. Since Eclipse ThreadX has a port for Windows, the projects in this chapter walk through the process of creating a project in Visual Studio that will run on Windows. Eclipse ThreadX is intended to be run on MCUs that are based on a different architecture than a PC. The Eclipse ThreadX port to Windows allows developers to learn

the features of Eclipse ThreadX and start application development while hardware is being developed. This parallel development approach would allow a GUI designer, with GUIX Studio, to write and test the application and user interface on Windows. Once the application is ready, the next step is to integrate it into the Eclipse ThreadX /MCU platform. GUIX Studio is an application available from the Microsoft Store. With GUIX Studio, you can create the graphic elements of the application, and the project is saved to an XML file with a .gxp extension. When finished, GUIX studio takes the graphic elements and creates C language files that can be integrated into the development tools for the target system. The graphics themselves are turned into pixel hex code.

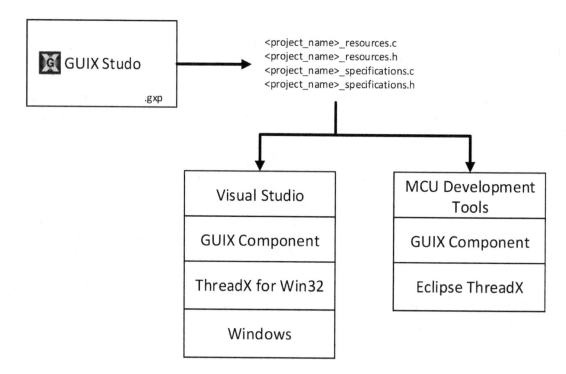

The power of portability allows the output from GUIX Studio to be used on different platforms with different processor architectures. The project for this chapter is to create a simple GUI in GUIX Studio, and then use Visual Studio to create a Win32 application using the output from GUIX Studio. The project is broken down into 4 computer activities.

12.2 Computer Activity 12.1: Tools Setup

Before we can write the application, three items need to be downloaded and installed.

1.   Open the Microsoft Store application.
2.   Search for "GUIX".
3.   Select Azure RTOS GUIX Studio from the results.

Note: Since the change in ownership, this application might have a new name.

4.   Click the install button to install the application.

The application will set up a workspace under C:\Azure_RTOS\GUIX-Studio-6.1. All projects will be stored in that folder.

5.   Start GUIX Store.
6.   You will be asked to download the GUIX repository (GUIX component) from GitHub. The interface doesn't work so click the Cancel button.
7.   Open a web browser
8.   Go to the following address: https://github.com/azure-rtos/guix .
9.   Click on the Code button and select "Download Zip".
10.   Save the Zip file to C:\Azure_RTOS\GUIX-Studio-6.1 .
11.   Once downloaded, unzip the guix-master.zip file.
12.   Rename the unzipped folder from guix-master to guix.

The contents of the guix folder look as follows:

IX-Studio-6.1 > guix

Name

cmake

common

docs

fonts

graphics

ports

samples

tutorials

CMakeLists.txt

CONTRIBUTING.md

LICENSE.txt

LICENSED-HARDWARE.txt

README.md

SECURITY.md

The common and ports folders contain the main source code for the GUIX component. Samples and tutorials provide different examples of how to set up the different widgets or to see how a full application is implemented. The samples and tutorials all come with Visual Studio solution files so you can run the sample on the desktop. Finally, graphics and fonts are resource files that are used in the projects. The C:\Azure_RTOS\GUIX-Studio-6.1\guix\ports\win32\build\vs_2019 folder contains libraries needed for the project.

13. In the web browser, go to the following address: https://github.com/azure-rtos/threadx .
14. Click on the Code button and select "Download Zip".
15. Save the Zip file to C:\Azure_RTOS\GUIX-Studio-6.1 .
16. Once downloaded, unzip the threadx-master.zip file.
17. Rename the unzipped folder from threadx-master to threadx.

The contents of the threadx folder look as follows:

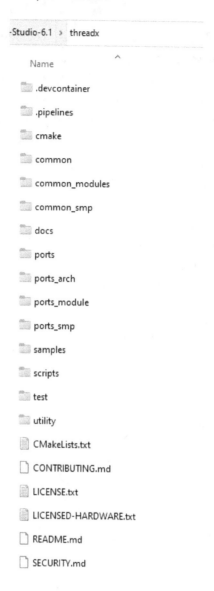

Three folders make up the common ThreadX source files. Four directories that cover different ports to different MCUs and MCU architecture types. The samples, test, and utility folders provide demo and diagnostic utilities. The C:\Azure_RTOS\GUIX-Studio-6.1\threadx\ports\win32\vs_2019 folder contains libraries needed for the project.

12.3 Computer Activity 12.2: Create a GUI using GUIX Studio

This simple demo will have a button and a text box. Each time the button is clicked, the text box will toggle between displaying one of two messages.

1. Open File Explorer.
2. Create a new folder under C:\Azure_RTOS\GUIX-Studio-6.1 called hello_world

Note: never use a dash (-) in a name.

3. Open GUIX Studio.
4. From the menu, select Project->New Project.
5. In the dialog box that appears enter the following:
    a. Project Name: hello_world .
    b. Project Path: C:\Azure_RTOS\GUIX-Studio-6.1\hello_world .

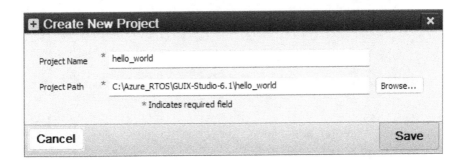

6. Click the Save button.
7. Another dialog appears.
8. In this dialog box, change the following:
    a. X resolution: 800
    b. Y resolution: 0
    c. Name: main_display
9. Click the Save button.

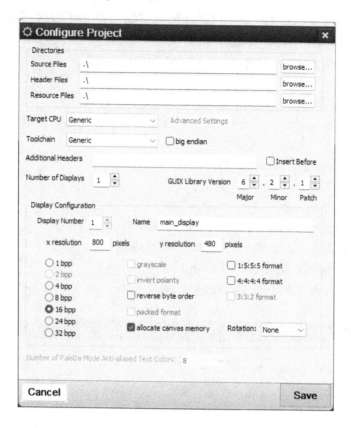

10. The project is created and a dialog appears with a simple help message. Click OK to close the dialog.

If you are familiar with creating Windows applications in Visual Studio, GUIX Studio development is very different. Controls are called Widgets. The main window for the applications is of the Widget Type: window. Each widget has a set of properties that can be configured. There are resources on the right for picture images, fonts, colors, and strings.

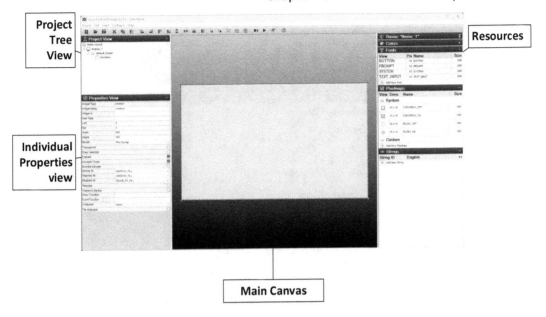

Project Tree View

Resources

Individual Properties view

Main Canvas

11. With the main canvas selected, in the Properties View change Widget Name to hello_world.

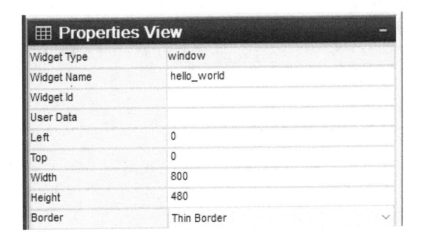

12. In the SH12-GUIX folder, there is a ClickMe.png source that will be used in the project. Expand the Pixelmaps in the resources on the right.

13. Click on Add New Pixelmap.

14. Open the ClickMe.png file. The resource is added to the available resources.

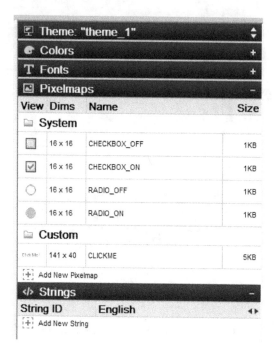

15. Right-click on the canvas.

16. A context menu appears. Select Insert->Button->Pixalmap Button.

17. A button is placed on the canvas. In the properties, set the following:

   a. Widget Id: ID_BUTTON_CLICKME

   b. Width: 141

   c. Height: 40

   d. Normal Pixelmap: CLICKME

Properties View		
Widget Type	pixelmap_button	
Widget Name	pixelmap_button	
Widget Id	ID_BUTTON_CLICKME	
User Data		
Left	325	
Top	302	
Width	141	
Height	40	
Border	No Border	∨
Transparent		☐
Draw Selected		☐
Enabled		☑
Accepts Focus		☑
Runtime Allocate		☐
Normal fill	BTN_LOWER	∨
Selected fill	BTN_UPPER	∨
Disabled fill	DISABLED_FILL	∨
Draw Function		
Event Function		
Pushed		☐
Toggle		☐
Radio		☐
Auto Repeat		☐
Normal Pixelmap	CLICKME	∨
Selected Pixelmap	None	∨
Disabled Pixelmap	None	∨
Horizontal Align	Left	∨
Vertical Align	Top	∨

18. Use the mouse to move the button to the low center of the canvas.
19. Right-click on the canvas.
20. From the context menu, select Insert->Text->Single Line Input
21. In the properties set the following:
    a. Width: 300
    b. Height: 40
    c. Text: Well, Hello There!
    d. Normal Text Color: Text:
22. Adjust the position of the Single Line Input box to the center of the upper half of the canvas.

23. Save the project.
24. Now, we need to generate the source code to create the application. From the menu, select Project->Generate All Output Files.
25. A dialog appears asking what to export. Keep the defaults and click the Generate button.

26. A dialog appears when the output files have been generated. Click the OK button.

27. Open File Explorer and go to the C:\Azure_RTOS\GUIX-Studio-6.1\hello-world folder. The C source code files have been generated for the project. We are ready to move to the next step to create an application in Visual Studio.

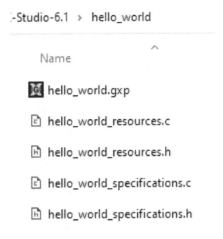

28. Close GUIX Studio.

12.4 Computer Activity 12.3: Building the Libraries

The GUIX and ThreadX libraries that are ported to Windows will be linked to our project in Visual Studio. The current libraries have been built with Visual Studio 2019. We will first rebuild the libraries with Visual Studio 2022 so they can easily be linked with the application.

1. Open Visual Studio.
2. Open the guix.sln file under C:\Azure_RTOS\GUIX-Studio-6.1\guix\ports\win32\build\vs_2019.
3. You may be asked to upgrade the solution. Click Ok.
4. Leave the build type set to debug x86 (Win32), and build the project.

The new gx.lib will be placed under the C:\Azure_RTOS\GUIX-Studio-6.1\guix\ports\win32\build\vs_2019 \Debug_GUIX_5_4_0_COMPATIBILITY folder.

5. Close the project.
6. Open the azure_rtos.sln file under C:\Azure_RTOS\GUIX-Studio-6.1\threadx\ports\win32\vs_2019\example_build. The project is actually building the library and a demo app. All we will need is the library in the end.
7. You may be asked to upgrade the solution. Click Ok.
8. Leave the build type set to debug x86 (Win32), and build the project.
9. The new tx.lib will be placed under the C:\Azure_RTOS\GUIX-Studio-6.1\threadx\ports\win32\vs_2019\example_build\tx\Debug folder.

Now, we want to update the libraries and needed .h files with the GUIX application.

10. Create a folder under C:\Azure_RTOS\GUIX-Studio-6.1\ called libraries.
11. Copy the tx.lib and gw.lib to the C:\Azure_RTOS\GUIX-Studio-6.1\libraries folder.
12. Copy the gx_api.h file, found under the C:\Azure_RTOS\GUIX-Studio-6.1\guix\common\inc folder, to the C:\Azure_RTOS\GUIX-Studio-6.1\libraries folder.
13. Copy the tx_api.h file, found under the C:\Azure_RTOS\GUIX-Studio-6.1\threadx\common\inc folder, to the C:\Azure_RTOS\GUIX-Studio-6.1\libraries folder.
14. Copy the tx_port.h file, found under the C:\Azure_RTOS\GUIX-Studio-6.1\threadx\ports\win32\vs_2019\inc folder, to the C:\Azure_RTOS\GUIX-Studio-6.1\libraries folder.
15. Copy the gx_port.h and gw_win32_display_driver.h files, found under the C:\Azure_RTOS\GUIX-Studio-6.1\guix\ports\win32\inc folder, to the C:\Azure_RTOS\GUIX-Studio-6.1\libraries folder.

The C:\Azure_RTOS\GUIX-Studio-6.1\libraries folder should look as follows:

X-Studio-6.1 › libraries

Name

gx.lib

gx_api.h

gx_port.h

gx_win32_display_driver.h

tx.lib

tx_api.h

tx_port.h

Putting the libraries and necessary .h files in a single location allows them to be re-used for other projects.

12.5 Computer Activity 12.4: Creating a Win32 Application

With the libraries created, the final step is to create the application in Visual Studio.

1. Create a new C Application project under C:\Azure_RTOS\GUIX-Studio-6.1\hello_world, called helloworld-Test.
2. Set the build type to Debug x86

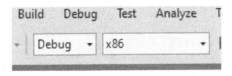

3. We need to import the source files from GUIX Studio. In Solution Explorer, right-click on Source Files and select Add-> Existing item... from the context menu.
4. Locate the hello_world_resources.c and hello_world_specification.c files and add them to the project.

5. In Solution Explorer, right-click on Header Files and select Add-> Existing item... from the context menu.
6. Locate the hello_world_resources.h and hello_world_specification.h files and add them to the project.
7. Right-click on Header Files and select Add-> Existing item... from the context menu.
8. Locate the gx_api.h, gx_port.h, gx_win32_display_driver.h, tx_port.h, and tx_api.h files and add them to the project.
9. Now we need to configure the properties of the project to point to the library resources. From the menu, select Project>Properties.
10. In the dialog, go down the tree on the left and expand C/C++.
11. Under General, click on the Additional Include Directories.
12. Hit the drop-down and click Edit.
13. A new dialog appears. Click on the new line folder icon and add C:\Azure_RTOS\GUIX-Studio-6.1\libraries to the list.
14. Create another new line and add C:\Azure_RTOS\GUIX-Studio-6.1\hello_world to the list.

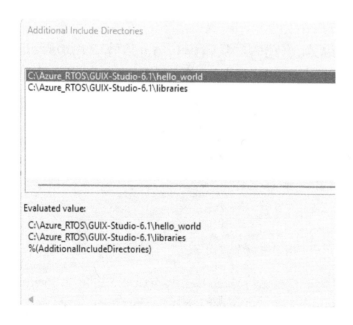

15. Click Ok.
16. Expand the Linker tree.

17. Under General, click on Additional Library Directories.

18. Hit the drop-down and click Edit.

19. A new dialog appears. Click on the new line folder icon and add C:\Azure_RTOS\GUIX-Studio-6.1\libraries to the list.

20. Click OK.

21. Under Input, click on Additional Dependencies.

22. Hit the drop-down and click Edit.

23. A new dialog appears. Enter gx.lib and tx.lib on separate lines.

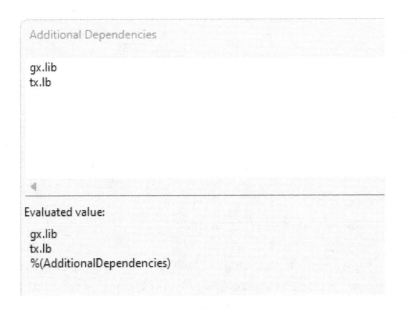

24. Click Ok.

25. Under All Options, change SubSystem to Windows (/SUBSYSTEM:WINDOWS). This change is necessary if this is going to be a Windows application.

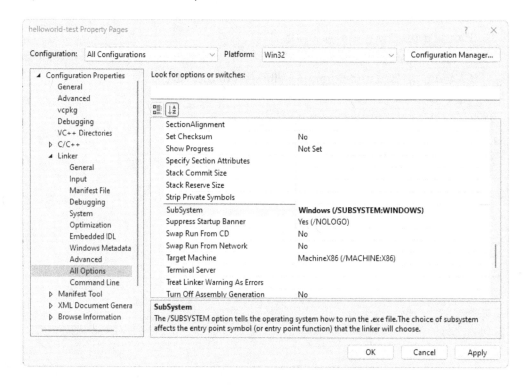

26. Click OK to close the properties dialog.

If you are familiar with developing Windows WPF or Form applications. Visual Studio has a lot of automation. In Visual Studio Designer, double-clicking on the control will open up a source code file where you can fill in the code for that control. For example, button controls have event handlers automatically created and all you have to do is write the code behind the button event. GUIX Studio only lets you create the GUI, and you have to add all the code behind the scenes a little later. There is a little more work to do to create the application. With that understanding, let's write the code, and then we can explain the details.

27. Open Source.c and enter the following:

```
#include <stdio.h>
#include "tx_api.h"
#include "gx_api.h"

/* Include GUIX resource and specification files for example. */
```

```c
#include "hello_world_resources.h"
#include "hello_world_specifications.h"

/* Define the new example thread control block and stack. */

TX_THREAD guix_thread;
UCHAR guix_thread_stack[4096];

/* Define the root window pointer. */

GX_WINDOW_ROOT* root_window;

/* Define function prototypes. */

VOID guix_thread_entry(ULONG thread_input);
UINT win32_graphics_driver_setup_565rgb(GX_DISPLAY* display);

int toggle = 0;

int main(int argc, char** argv)
{
    /* Enter the ThreadX kernel. */
    tx_kernel_enter();
    return(0);
}

VOID tx_application_define(void* first_unused_memory)
{
    /* Create the new example thread. */
    tx_thread_create(&guix_thread, "GUIX Thread", guix_thread_entry,
0, guix_thread_stack, sizeof(guix_thread_stack), 1, 1,
TX_NO_TIME_SLICE, TX_AUTO_START);
}

VOID guix_thread_entry(ULONG thread_input)
{

    /* Initialize the GUIX library */
    gx_system_initialize();

    /* Configure the main display. */
    gx_studio_display_configure(MAIN_DISPLAY,  /* Display to
configure*/
        win32_graphics_driver_setup_565rgb, /* Driver to use */
        LANGUAGE_ENGLISH,                   /* Language to install */
        0,         /* Theme to install */
        &root_window);                      /* Root window pointer */
```

```
    /* Create the screen - attached to root window. */
    gx_studio_named_widget_create("hello_world",
(GX_WIDGET*)root_window, GX_NULL);

    /* Show the root window to make it visible. */
    gx_widget_show(root_window);

    /* Let GUIX run. */
    gx_system_start();

}

UINT app_event_handler(GX_WINDOW* window, GX_EVENT* event_ptr)
{

    switch (event_ptr->gx_event_type)
    {

    case GX_SIGNAL(ID_BUTTON_CLICKME, GX_EVENT_CLICKED):

        if (toggle == 0) {
            /* Clear input buffer.  */

gx_single_line_text_input_buffer_clear(&hello_world.hello_world_text_
input);

gx_single_line_text_input_character_insert(&hello_world.hello_world_t
ext_input, "Welcome to", 10);

            toggle = 1;
        }
        else
        {

gx_single_line_text_input_buffer_clear(&hello_world.hello_world_text_
input);

gx_single_line_text_input_character_insert(&hello_world.hello_world_t
ext_input, "Eclipse ThreadX ", 10);
            toggle = 0;
        }
        break;

    default:

        return gx_window_event_process(window, event_ptr);
    }
```

```
    return 0;
}
```

28. Save the file.

The source code is based on the example source code from the Eclipse ThreadX documentation. When the application launches, the user can click on the button and the message in the text box will toggle between 2 messages. Starting from the top of the source code listing are the header files to include, which consist of the two Eclipse ThreadX component libraries and the header files from the GUIX Studio output. A thread, thread stack, and GUIX root windows are defined. Function prototypes for the GUIX thread and the display driver are defined. The toggle variable is declared.

The main function makes a call to start the ThreadX kernel, which is part of tx.lib. In the process of starting the kernel, the tx_applicaiton_define function is called, which creates a thread to start the GUIX window and display the GUI we created in GUIX Studio.

The guix_thread_entry is the thread function. A call is made to start the GUIX library, and then the gx_studio_display_configure is called to associate the MAIN_DISPLAY and the video driver that is part of the GUIX library (gx.lib) to the root_window.

In Computer Activity 12.2, we set the display name to "main_display" and the resolution to 800x480 in GUIX Designer. The output from file hello_world_resources.h has defined that information as follows:

```
/* Display and theme definitions
*/

#define MAIN_DISPLAY 0
#define MAIN_DISPLAY_COLOR_FORMAT GX_COLOR_FORMAT_565RGB
#define MAIN_DISPLAY_X_RESOLUTION 800
#define MAIN_DISPLAY_Y_RESOLUTION 480
#define MAIN_DISPLAY_THEME_1 0
#define MAIN_DISPLAY_THEME_TABLE_SIZE 1
```

The MAIN_DSPLAY is used in the gx_studio_display_configure function as the first and only display. The gx.lib source code includes several driver types. The win32_graphics_driver_setup_565rgb was chosen to match the color format support. The

225

gx_studio_named_widget_create function makes the final connection between the hello_world canvas widget and the root_window. The last two function calls are to show the root_window and start GUIX.

The code has primarily focused on creating and starting the GUI. Events have to be addressed separately. The app_event_handler function handles all the events that take place with the application. There are 63 widget events defined in gx_api.h. When the GUI is launched, the app_event_handler is called. Since there are no matching cases in the Switch-case statement, the gx_window_event_process is called as the return. When the ClickMe button is clicked, the app_event_handler is called, the case is matched as a GX_EVENT_CLICKD event and the text box changes messages based on the toggle value. The button was named ID_BUTTON_CLICKME in GUIX Studio, and the widget ID is defined in the GUIX Studio output file hellow_world_specifications:

```
/* Define widget ids
*/

#define ID_BUTTON_CLICKME 1
```

The &hello_world.hello_world_text_input is the address of the text box. The hello_world_specification.h links the widgets to the hello_world application.

```
/* Declare top-level control blocks
*/

typedef struct HELLO_WORLD_CONTROL_BLOCK_STRUCT
{
    GX_WINDOW_MEMBERS_DECLARE
    GX_PIXELMAP_BUTTON hello_world_pixelmap_button;
    GX_SINGLE_LINE_TEXT_INPUT hello_world_text_input;
} HELLO_WORLD_CONTROL_BLOCK;


/* extern statically defined control blocks
*/

#ifndef GUIX_STUDIO_GENERATED_FILE
extern HELLO_WORLD_CONTROL_BLOCK hello_world;
#endif
```

You can see that the names and settings defined in GUIX Studio translate into code and are called in the main application. For the event handler to be called, a link has to be made to the hello_world widget.

29. Open hello_world_specification.h
30. Around line 113 after the comment to defined event process functions, etc. add the following to declare the function:

```
/* Declare event process functions, draw functions, and callback
functions      */
UINT app_event_handler(GX_WINDOW* window, GX_EVENT* event_ptr);
```

31. Save the file.
32. Open hello_world_specification.c, around line 194 in the GX_CONST GX_STUDIO_WIDGET hello_world_define, replace the GX_NULL for event function override with the app_event_handler:

```
GX_CONST GX_STUDIO_WIDGET hello_world_define =
{
    "hello_world",
    GX_TYPE_WINDOW,                          /* widget type        */
    GX_ID_NONE,                              /* widget id          */
    #if defined(GX_WIDGET_USER_DATA)
    0,                                       /* user data          */
    #endif
    GX_STYLE_BORDER_THIN|GX_STYLE_ENABLED,   /* style flags        */
    GX_STATUS_ACCEPTS_FOCUS,                 /* status flags       */
    sizeof(HELLO_WORLD_CONTROL_BLOCK),       /* control block size */
    GX_COLOR_ID_WINDOW_FILL,                 /* normal color id    */
    GX_COLOR_ID_WINDOW_FILL,                 /* selected color id  */
    GX_COLOR_ID_DISABLED_FILL,               /* disabled color id  */
    gx_studio_window_create,                 /* create function    */
    GX_NULL,                                 /* drawing function
override        */
```

```
    (UINT(*)(GX_WIDGET*, GX_EVENT*)) app_event_handler,      /* event
function override          */
    {0, 0, 799, 479},                            /* widget size
*/
    GX_NULL,                                     /* next widget
*/
    &hello_world_pixelmap_button_define,         /* child widget
*/
    0,                                           /* control block
*/
    (void *) &hello_world_properties             /* extended properties
*/
};
```

33. Save the file.
34. Build the application, and correct any errors.
35. Start a debug session. The application will start up and show the initial screen:

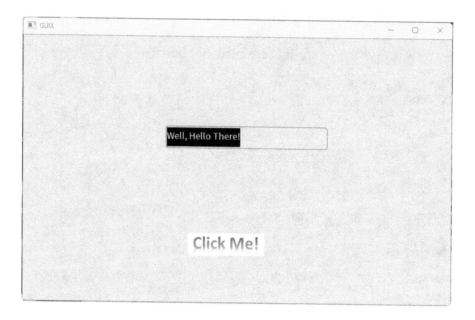

36. Click the ClickMe button a few times to toggle the Welcome to Eclipse ThreadX message.
37. When finished stop the debug session.

12.6 Summary and Book Conclusion

You can create a small GUIX application that runs in Windows. The port to Windows allows you to test ThreadX features without the requirement for an ARM MCU platform. The port to Windows is only for simulation. There is no loader or port available to boot ThreadX on a PC, so you cannot take advantage of running ThreadX on a PC without Windows.

The last three chapters have introduced three of the main investments Microsoft has made in supporting the MCU market. Python, Perl, C#, and Java are popular language, but all of these MCU investments require C language developers. All the C programming concepts discussed in this book apply universally. Of course, an application designed to use a platform's custom libraries will only run on that platform. Please keep an eye out for our other publications.

In today's, programming environment, writing Windows applications in C might seem archaic. The Computer Activities are focused on teaching the concepts of C programming. There are some unique applications that can be developed in C for Windows. We recommend the book *Hands-On Network Programming with C* by Lewis Van Winkle if you are interested in diving deeper into network programming using the C language. The ISBN is listed in the bibliography.

A Bibliography

Besides the online help files, various books, articles, websites, and presentations were used in the development of this book.

A.1 Books:

The C Programming Language, Second Edition, Brian W. Kernighan and Dennis M. Ritchie, Prentice Hall, 1988, ISBN: 0-13-110362-8

The C Answer Book, Second Edition, Clovis L. Tondo and Scott E. Gimpel, Prentice Hall, 1988, ISBN: 0-13-109653-2

The Standard C Library, P.J. Plauger, Prentice Hall, 1992, ISBN: 0-13-131509-9

The CERT® C Coding Standard, Second Edition, Robert C. Seacord, Addison-Wesley, 2014, ISBN: 978-0-321-98404-3

Technical C Programming, Vincent Kassab, Prentice Hall, 1989, ISBN: 0-13-898339-9

Java and Eclipse for Computer Science, Sean D. Liming and John R. Malin, Annabooks, LLC., 2018: ISBN: 978-0-9911887-3-4

Extreme C, Kamran Amini, Packt, 2019, ISBN: 978-1-78934-362-5

Hands-On Network Programming with C, Lewis Van Winkle, Packt, 2019, ISBN: 978-1-78934-986-3

Using C, Lee Atkinson and Mark Atkinson, Que, 1990, ISBN: 0-88022-571-8

Appendix A - Bibliography

C Programmer's Toolkit, Second Edition, Jack Purdum, Que, 1992, ISBN: 0-88022-788-5

A.2 Websites:

- C standard library - Wikipedia

- The Current C Programming Language Standard - ISO/IEC 9899:2018 (C18) - ANSI Blog

- Memory Layout of a C Program | Hack The Developer.com

- C library function - free() (tutorialspoint.com)

- Overview of Azure IoT device SDK options | Microsoft Learn

- C SDK and Embedded C SDK usage scenarios | Microsoft Learn

- https://www.geeksforgeeks.org/types-of-recursions/

B INDEX